Understanding Digital Terrestrial Broadcasting

For a listing of recent titles in the *Artech House
Digital Audio and Video Library,* turn to the back of this book.

Understanding Digital Terrestrial Broadcasting

Seamus O'Leary

Artech House
Boston • London
www.artechhouse.com

Library of Congress Cataloging-in-Publication Data

O'Leary, Seamus.
 Understanding digital terrestrial broadcasting / Seamus O'Leary.
  p. cm. — (Artech House digital audio and video library)
 Includes bibliographical references and index.
 ISBN 1-58053-063-X (alk. paper)
 1. Digital communications. 2. Television broadcasting. 3. Digital audio
  broadcasting.
 I. Title. II. Series.
TK5103.7.O49 2000
621.384—dc21   00-040624
                                                                CIP

British Library Cataloguing in Publication Data

O'Leary, Seamus
 Understanding digital terrestrial broadcasting. — (Artech House digital audio and video
 library)
 1. Digital television 2. Digital audio broadcasting
 I. Title
 621.3'88

 ISBN 1-58053-063-X

Cover and text design by Darrell Judd

© 2000 ARTECH HOUSE, INC.
685 Canton Street
Norwood, MA 02062

International Standard Book Number: 1-58053-063-X
Library of Congress Catalog Card Number: 00-040624

10 9 8 7 6 5 4 3 2 1

This book is dedicated to the memory of Una and Denis McLoughlin,

Go raibh leaba i measc na naoimh agaibh i gcónaí.
(May you always have a bed amongst the Saints.)

Contents

Preface

This book is intended for use by engineers, technicians, students, and anyone interested in acquiring a working knowledge of the principles behind digital television and digital radio broadcasting. It assumes that the reader already has some knowledge of modern color television engineering. It treats in detail the European digital television standards, however, as it describes the universally accepted digital encoding practices, it will have general appeal. It also details the digital broadcasting systems adopted by the United States and other countries. The book begins in Chapter 1 with a general overview of the recent developments within digital television (DTV), and a review of some of the more important organizations working in this field. It also introduces the concepts of compression and redundancy, which are used in subsequent chapters dealing with encoding within studios.

It is intended to give the reader an in-depth understanding of the more important issues related to digital television engineering. The subject matter is backed up by the usage of references to more detailed information sources. Also, the topics are supported by the usage of diagrams and schematics, which give functional descriptions and help to illustrate the principles. It is not the purpose of the book to illustrate digital television techniques mathematically and hence this has been kept to a minimum. It is clear that at the time of writing this book great advances in digital signal processing (DSP) are having a profound effect in driving this technology forward. This will also yield improvements in equipment performance, and with very large scale integration (VLSI) of chip circuitry, the cost of this equipment will be reduced.

At the time of the writing of this book, digital television standards are being written by relevant standards bodies and adopted by different countries. This book makes reference to current developments and describes the more important variants that are in the process of adoption. As the most common broadcasting method is terrestrial broadcasting, particular emphasis is placed on terrestrial engineering matters in Chapter 8. Technologies applicable to other digital transmission media such as cable, satellite and microwave multipoint distribution systems (MMDS) are described in some detail in Chapter 6. There is much commonality between different digital transmission media such as terrestrial, MMDS, cable, and satellite, and it is generally accepted that high power terrestrial systems place the greatest constraints on equipment performance and engineering, hence the reason for the emphasis this book places on digital terrestrial television (DTT).

This book places emphasis on transmission aspects of digital television (DTV), including modulators and demodulators, the Digital Video Broadcasting standard for terrestrial Television (DVB-T), and the networks that are used to distribute and broadcast DTV signals. Chapter 14 deals with mobile broadcasting applications of DTV including mobile news gathering, and outside broadcasting. Single frequency networks (SFN) are about to be deployed for DTV in several countries. The more important aspects of these new networks are discussed in Chapter 10 together with multifrequency networks (MFN).

Details on the coverage expected from transmitters and the requirements for receivers in terms of carrier to noise ratios, and field strengths are given in Chapter 11. The more common tests and measurements made on DTV signals are detailed in Chapter 9, together with a discussion on set top boxes, and the software (middleware) needed to run applications, that is, the application programming interface.

Common scenarios of establishing DTV systems in parallel with analog TV are detailed and the particular challenges that these pose are discussed.

Chapter 13 discusses digital audio broadcasting (DAB) and other related variants of this technology. It describes some of the audio encoding standards within the MPEG family of digital compression standards. It also describes the emerging digital radio standards for medium wave and long wave radio broadcasts.

It is my opinion that the next big challenge facing the broadcasting and telecommunications sector will be co-existence within a digital world. Interactivity and the implications this will have for all network providers will have to be addressed. However, these topics are beyond the scope of this book.

Acknowledgements

I am very grateful to many people who have had a major influence on the conception and fruition of this book. I will now attempt to acknowledge their inputs. I would like to thank my ex-colleagues in North West Labs Ltd., and Cable Management Ireland for introducing me to the world of digital television, MMDS, cable, and the DVB project. I would like to thank Mel Healy for his support and enthusiasm. Also I wish to thank Dr. Ronan O'Dowd of UCD and Dr. David MacDonald of Optronics Ireland for developing my research methodologies.

I would like to thank all of those experts within the various consortiums that I have participated in at a European level for valuable advice and support. I am particularly grateful to the BBC research and development department, and the members of the EU projects DIMMP, VALIDATE, HD-SAT, BRITEUR, and DIAMOND.

My gratitude goes to the staff in Artech House for helping me with this project, and for their patience. I would like to thank Arch Luther for valuable and constructive reviews of the material.

I would like to thank the group of professional engineers that I work with for providing me with a stimulating and challenging environment on a daily basis. I am particularly indebted to Peter Branagan of RTE not only for reviewing and approving this manuscript for publication, but also for providing a clear vision of tomorrow's digital broadcasting environment. My colleagues, Malachy Donohoe, Brendan Crinion, Brian Wynne, Paul Myhal, Keith McInerney, Redmond Coady, Peter Rogers, David Sherwood, and Joe Kavanagh, have all helped shape my understanding of this technology, and for that I am very grateful.

I would like to thank my parents, Gertie and Paddy O'Leary for their support in this personal project. Also I am indebted to my family members, Padraig and Eamon, and my friends for their enthusiasm. In closing, I want to thank the most important person in the writing of this book, my wife Siobhan. This book would not have been possible without her unselfish support, and understanding, during the prolonged and difficult writing of this book. She also somehow, managed to keep my baby daughter, Niamh, distracted. Niamh has attempted in her own ways, many times, to contribute to the writing of this book, and I will always be grateful for her distractions.

CHAPTER

1

Contents

Current Situation

1.1 Introduction

At present there is huge interest in digital television (DTV). As we enter the twenty-first century, digital television is considered an integral part of the information super-highway that is being built for the new millennium. This is because digital television can deliver vast amounts of information at very low cost to the maximum number of viewers, it can now be fully integrated into completely digital transmission networks, and it can be packaged as never before.

Digital television can deliver more programs than traditional analog television over any transmission medium. This is because digital information can be manipulated and treated in ways never possible with analog television. It is easy to store digital images on computers and discs and play them continuously over digital networks without signal degradation. Images can be edited and enhanced, compressed and stored, transmitted and printed. By representing pictures as binary digits (0's and 1's) digital television is extremely flexible in how it treats information. Television signals, which in analog format require dedicated circuits, can in digital format be mixed (integrated) with telephone conversations and computer data and then transmitted over telecommunication networks to distant broadcasting sites. Programs can be stored on computer hard discs and retrieved instantly for broadcast to a single viewer on demand. The delivery of multimedia material (audio, video, and data) in digital format to the consumer creates the opportunity to store content using inexpensive personal computer-based technology. A computer hard disc can store a movie and retrieve and manipulate it in new ways. Clearly these developments represent a revolutionary change to traditional analog broadcasting.

To try and overcome implementation problems and ensure an orderly adoption of standards many international organizations have been working to solve engineering problems and set frameworks and standards for digital television implementation.

1.2 Historical Background and Future Trends in Digital Television

The current analog television systems are based on the National Television Systems Committee (NTSC) system developed in the United States in 1953. This system is used throughout America, Canada, Mexico, and Japan and in many parts of South America and Korea since 1954. The phase alternation line (PAL) system is a variant of the NTSC system and is used extensively throughout Europe, Australia, and the Far East in slightly different formats. There are other systems in use including SECAM and the satellite based MAC standards. These standards built barriers between the transmission of television services internationally. However all of these analog standards are set to be replaced over the coming decade by digital transmission standards which will be less restrictive in transmission and capable of delivery of new services including information channels carrying internet services and subscription programming.

Most current color receivers use 4:3 aspect ratio only, while some widescreen receivers can display 16:9 aspect ratio. However, as part of the move to digital television many broadcasters will consider the move to widescreen 16:9. It will be possible to offer digital viewers HDTV (high definition television) from possibly widescreen sources. It will be possible to offer existing (4:3) viewers 16:9 programs in 14:9 (half letterbox) format. New all digital processing chips are capable of delivery of low cost consumer receiver units, which will demodulate and decode digital television signals.

In the analog TV broadcasting era organizations often owned the terrestrial network transmission facilities. For mainly historical reasons, such as the large capital cost of network equipment, many terrestrial networks were state or government owned. This lack of access to the airwaves for new competitors, gave protection from competition to the existing network operators. Such was the effectiveness of this barrier to market entry that for many years very few broadcasters were involved in terrestrial systems. However with the advent of digital technologies there will be plenty of bandwidth available for other program providers and hence opportunities for program makers. There will be further separation between program makers and network providers, with traditional broadcasting practices being replaced by more flexible and competitive methods of operation. With competition from telecommunication network operators in the provision of DTV on competing media such as cable and terrestrial systems, the industry is currently undergoing enormous change.

1.3 Global Developments in Digital Television

The development of DTV in the United States, Europe, and Japan has occurred at slightly different times. The main DTV standardization organizations of these three regions have been influenced somewhat by other related developments, such as the work of the Moving Pictures Expert Group (MPEG) toward video and audio encoding standards, system information, and multiplexing standards. These standardization organizations have worked to develop broadcast modulation standards suitable for the type of media and channel bandwidth already in use within that particular region. As a result there are differences between the modulations used, for example, in the United States and Europe. Some of the more important milestones in the development of DTV are shown in Table 1.1.

1.3.1 DTV in the United States

In 1987 the Federal Communication Commission (FCC) initiated a process to select a suitable High Definition Television (HDTV) standard for the United States that would be compatible with the existing analog television standard (NTSC). By 1992 four proposals were short-listed, and by 1993 agreement was reached by the four proposing consortia to form a Grand Alliance (GA) to complete development of the standard [1].

The GA has specified a Dolby standard called AC-3 for multichannel audio source encoding, and has specified MPEG's standard known as MPEG-2 for video source encoding, system information, and multiplexing. The GA has also specified an 8-level vestigial sideband modulation (8 VSB) for terrestrial broadcasting with a payload of 19.28 Mbps in a 6 MHz bandwidth broadcast channel. The GA has also specified a system for cable television systems.

1.3.2 DTV in Europe

In Europe many projects were undertaken in the early 1990s to specify a HDTV standard and with the help of the German government, a European Launching Group (ELG) was formed in 1992 that invited participation from interested organizations in Europe. With the success of the ELG in 1993 approximately 84 broadcasters, standards bodies, telecommunications companies, manufacturers, and other organizations formed the Digital Video Broadcasting project (DVB) by signing a memorandum of understanding. The project has grown in membership steadily since that date.

Based in the European Broadcasting Union (EBU) headquarters in Geneva and supported by the European Commission, the DVB project has developed standards for broadcasting on different media such as cable, satellite, and terrestrial channels. The DVB project has specified MPEG-2 as the source encoding standard for audio, video, as well as system information, and multiplexing. The DVB project has specified Coded Orthogonal Frequency Division Multiplexing (COFDM) as the terrestrial broadcast channel modulation standard, and it is referred to as the DVB-T standard.

1.3.3 DTV in Japan

Japan officially started the development of DTV in 1994, and the Japanese Ministry of Post and Telecommunications (MPT) has coordinated the work. The Japanese have adopted the MPEG-2 system for source encoding and system information and have established the Japanese Digital Broadcasting Experts Group (DiBEG) to formulate a strategy for digital broadcasting on various transmission media. The Japanese are evaluating a variant of the COFDM system for terrestrial broadcasting known as integrated services digital broadcasting-terrestrial, (ISDB-T).

Table 1.1

Some Milestones in the Development of DTV

Date	Milestone
1990	First proposal for a digital terrestrial HDTV system from General Instrument
1991	Scandinavian HD-DIVINE project for terrestrial HDTV
1992	ELG formed in Europe
1993	GA formed in the United States
1993	DVB project starts
1994	Japanese MPT founds the Digital Broadcasting Development Office
1994	DVB produce European Common Antenna TV (CATV) standard
1994	DVB produce Direct To Home (DTH) satellite standard
1995	ATSC DTV standard A/53 in the United States
1995	DVB produce SMATV (Satellite Master Antenna Television) standard
1995	ATSC digital audio compression (AC/3) standard A/52
1995	DVB produce common scrambling algorithm for conditional access (CA)
1996	DVB produce common interface for CA
1996	Canal Satellite becomes Europe's first public digital broadcaster via satellite
1996	First public digital satellite broadcasting in Japan by PerfecTV
1997	DVB specify standard for digital MMDS systems (microwave multipoint distribution systems)
1997	DVB produce European standard for terrestrial systems
1998	Digital terrestrial services launched in the United Kingdom
1998	Digital terrestrial transmissions begin in the United States
1998	Australia adopts the DVB terrestrial system

This standard, expected in the early years of the new millennium, should have enhanced mobility features over the existing DVB-T standard.

1.4 Digital Television Organizations

Many organizations are presently working on standards for digital television and making recommendations to world standard bodies. These standards cover areas including encoding, decoding, modulation, framing, frequency coordination, encryption, conditional access, transport, and so forth. Some of the more important groups include:

MPEG	(Moving Pictures Expert Group)
MHEG	(Multimedia/Hypermedia Expert Group)
DVB	(Digital Video Broadcasting) project
DAVIC	(Digital Audio-Visual Council)
EBU	(European Broadcasting Union)
ITU	(International Telecommunication Union)
ETSI	(European Telecommunications Standards Institute)
ANSI	(American National Standards Institute)
ATSC	(Advanced Television System Committee)
IEC	(International Electrotechnical Commission)
ISO	(International Organization for Standardization)
Digitag	(Digital Terrestrial Television Action Group)
DTG	(U.K. Digital TV group)
CENELEC	(European Committee for standardization)
DiBEG	(Japanese Digital Broadcasting Experts Group)

MPEG—The Moving Picture Experts Group is a working group of ISO/IEC in charge of the development of international standards for compression, decompression, processing, and coded representation of moving pictures, audio and their combination. It is a subgroup of a joint ISO/IEC technical committee that is standardizing information technology related equipment, (refer to JTC 1 below).

MPEG has produced standards including:

MPEG-1	a standard for the storage and retrieval of moving pictures and audio on digital storage media.
MPEG-2	a standard for digital television broadcasting.

Two more standards are under development:

MPEG-4	a standard for multimedia applications.
MPEG-7	an Audio-Visual content representation standard for fast information searching and retrieval.

MPEG usually holds three meetings a year. These comprise plenary meetings and subgroup meetings on requirements, systems, video, audio, test, implementation, liaison, and other topics. MPEG meetings are typically attended by some 300 experts from more than 20 countries.

MHEG—The Multimedia/Hypermedia Expert Group is another working group under the same subcommittee that features MPEG. MHEG targets coding of multimedia and hypermedia information, and defines an interchange format for composite multimedia contents. The defined MHEG format encapsulates a multimedia document, as communication takes place in a specific data structure.

DVB—The Digital Video Broadcasting (DVB) project comprises a group of more than 200 organizations from more than 25 different countries working together to establish the technical framework for the introduction of digital broadcasting systems. DVB has already established many European standards that have been ratified by ETSI, CENELEC, and the ITU, these include:

DVB-T   Terrestrial television transmission standard

DVB-C   Cable television transmission standard

DVB-S   Satellite television transmission standard

DVB-SI   Specification for service information

DVB-CS   SMATV (Satellite Master Antenna Television)
           transmission standard

DVB-TXT   Teletext transport specification

DVB is working on other standards including:

DVB-TRC   The digital terrestrial return channel

DVB-MHP   The digital multimedia home platform

DVB-MC   The digital Microwave Multipoint Distribution
           System (MMDS) below 10 GHz (the more commonly
           used system is at 2.5–2.7 GHz)

DVB-MS   The digital microwave multipoint distribution
           system (MMDS) above 10 GHz

DVB-CI   The DVB common interface for use in conditional
           access (CA)

The DVB project meets a number of times each year in the EBU Headquarters in Geneva and is working on all aspects of transmission.

DAVIC—The Digital Audio-Visual Council was created in 1994 and covers an extremely wide field. DAVIC is a nonprofit organization based in Switzerland, with a membership of over 175 companies from more than 25 countries. DAVIC seeks to provide end-to-end interoperability for the use of digital images and sound across countries and between applications. There is much liaison between DVB and DAVIC. DAVIC provides specifications of open interfaces and protocols for digital services and applications. The specifications are used as tools to aid setting interfaces for video on demand (VoD), sever to set-top communication and other equipment interfaces. Multimedia applications and interactivity are covered by DAVIC. DAVIC is also looking at Internet applications and standards conversion. The most recent focus of DAVIC has been on the idea of "TV anytime/TV anywhere," where wanted programming is searched for, then identified, and finally downloaded to a local receiver for consumption at a convenient time. DAVIC has recently decided to conclude its existence.

EBU—The European Broadcasting Union is an independent association of national broadcasters. It is nongovernmental and noncommercial. Non-European broadcasters can join the EBU as associate members. It supports the activities of the DVB project and DigiTAG and contributes to the work of other standards bodies such as CENELEC, ETSI, ITU, and IEC. The EBU establishes and publishes recommendations and standards, which are often considered by the ITU and/or the IEC to be turned into world standards. It also runs the Eurovision satellite network on Eutelsat-F4, which provides high quality contribution "feeds" to national broadcasters from approximately 50 ground stations. It uses MPEG-2, 4:2:2 encoding on the network that it operates.

ITU—The International Telecommunication Union headquartered in Geneva, Switzerland is an international organization within which governments and the private sector coordinate global telecom networks and services. It is an agency under the United Nations (UN). The ITU is responsible for frequency spectrum management and is a leading publisher of telecommunication technology, regulatory and standards information. It has activities in standardization of digital television through the ITU-R (Radiocommunication Sector) and the ITU-T (Telecommunication Standardization Sector). The ITU-T recommendation J.81 for the transmission of video at 34 Mbps on Telecommunications networks is identical to the ETSI standard (ETS 300 174). Within the ITU-R two working parties are active in the field of DTV, working party 11/3 in the field of digital terrestrial systems, and working party 10-11/S in the field of satellite systems. The ITU is the most important standardization organization at a global level in the field of telecommunications.

ETSI and ANSI—The European Telecommunications Standards Institute (ETSI) and the American National Standards Institute (ANSI) have carried out a considerable amount of work to specify the interconnection of video transmission circuits to telecommunication equipment. This has resulted in standards such as ETS 300 174 (equal to ITU-T Rec. J.81) and ANSI standard T1.802.01 for contribution and high quality distribution of digital video. Both standards specify only one video channel per bit stream. They both specify encoding, multiplexing, scrambling, and network adaptation to allow direct interconnection to telecommunications equipment.

ETSI was formed in 1988 to aid in establishing a single market in Europe by setting telecommunication standards. ETSI produces standards through technical committees to interconnect public and private networks.

ANSI is a voluntary and privately funded business standards group in America. It has over 1,500 members from U.S. and international companies, professional institutes, consumer groups, and government agencies, and is the only U.S. member of ISO.

Standard ETSI codecs (encoders/decoders) are used to connect broadcasters to and from telecommunication networks. The codec will encode the incoming signal and compress it. The digitized signal is then multiplexed, scrambled, and transmitted at approximately 34 bps.

ANSI codecs are similar to ETSI codices except for the audio interface (AS/BE) and SMITE machine control. The final bit rate is compatible with U.S. telecom networks working at a bit rate of 45 Mbps.

ATSC—The Advanced Television System Committee was formed in the United States to establish voluntary technical standards for advanced television systems, including digital high definition television (HDTV). The ATSC is supported by its members, who are themselves subject to certain qualification requirements. On December 24, 1996, the United States Federal Communications Commission (FCC) adopted the major elements of the ATSC Digital Television Standard (A/53) for the United States next generation of broadcast television. Under the decision, the video and audio compression, the packetized data transport structure, and the modulation and transmission system specified in the ATSC standard are mandated by the Commission for use by terrestrial broadcasters. However, the specific video formats to be used for digital broadcast television will be the subject of voluntary industry standards.

IEC—The International Electrotechnical Commission is responsible for electrical and electronic standardization of equipment and components.

ISO—The International Organization for standardization is a nongovernmental federation of national standards bodies that sets standards in industry.

Both ISO and IEC work at a global level on the standardization of consumer and industrial equipment. They have established many Joint Technical Committees to work on overlapping areas of interest.

JTC 1—Due to the overlap between ISO and IEC in the field of information technology a Joint Technical Committee 1 (JTC 1) was established. JTC 1 is responsible for standardising information technology related equipment. A JTC 1 sub-group was formed to standardize the encoding of digital video and audio compression equipment. This sub-group is called the Moving Pictures Expert Group (MPEG) and we shall refer to the MPEG family of standards throughout this book.

DigiTAG—The Digital Terrestrial Television Action Group was launched in 1996 to create a framework for the harmonious introduction of digital terrestrial television using the DVB-T specification. It has over 50 members from 18 countries and is administered through the European Broadcasting Union (EBU).

DTG—U.K. Digital TV Group, this is a U.K. and Irish organization open to members of the DVB project who are interested in the development of digital terrestrial television and who are undertaking research contribution to the deployment of such a technology. It has produced recommendations and technical proposals to the relevant U.K. government authorities. The DTG group is growing in membership and now offers overseas affiliate membership status to foreign organizations interested in DTV. It produces a newsletter and technical documents for the broadcasting community.

CENELEC is the European Committee for Electrotechnical Standardization. It was set up in 1973 as a nonprofit-making organization under Belgian Law. It has been officially recognized as the European standards organization in its field by the European Commission in Directive 83/189 EEC. Its members have been working together in the interests of European harmonization, alongside the European Economic Community. CENELEC works with technical experts from 19 EC and EFTA countries to publish standards for the European market. The organization incorporates technical committees in the field of radio receivers, television, conditional access, and cable distribution systems. It is therefore a suitable organization to contribute to the standardization of DTV.

1.5 The Need for Compression

Present day television is being generated, transmitted, and viewed in various forms. Many production studios and cameras are totally digital, with digital representations of the colors that the camera outputs (red, green, and blue). More and more transmission infrastructure is going digital, with the television signals transported over telecommunication networks in digital formats which are compatible with telephony signals. At, for example, an analog terrestrial TV transmitter site, the digital signals are converted to an analog transmission standard such as PAL or NTSC and then these composite analog signals are transmitted over the air to domestic receivers where they are received off-air and displayed on analog TV sets. As the reader may understand this involves many changes in the format of the TV signal from analog to digital and back again depending on the particular network and studio architecture used. Maintaining a signal in digital format is to be preferred as it allows for robust storage and transmission with constant quality. The interfaces to other digital equipment are easier to implement and efficient use of bandwidth can be managed. By minimizing the number of format changes (especially from analog composite format to digital format) of the information signal, the amount of coding artifacts that are introduced to the signal can be kept to a minimum. This is to be desired if the quality of the signal is to be maintained from the studio to the home.

1.5.1 Redundancy and Entropy

The key to widespread usage of digital audio-visual technologies is the correct usage of compression. When an analog video signal is linearly sampled the amount of digital data generated is enormous. Compression is a flexible technology because the amount of compression used, and the degree of encoding complexity can be tailored to suit the application. Video signals are known to contain redundancy because the images contain areas that are similar, and images do not change completely from frame to frame. On the other hand, entropy may also be associated with a video signal. Entropy is a measure of unpredictability in data. The actual information in a video signal has high entropy as it is unpredictable. The remainder of the signal is predictable and therefore has low entropy.

It is important to use compression techniques that suit the application. For the production and processing of material it is important to ensure that the material is suitable for editing at a later stage. As a result compression techniques should ideally remove redundancy only within the same field, which is termed intracoding (see Section 1.6.1). No advantage is taken of the redundancy from field to field as this would compromise editing. However, for distribution and broadcasting purposes higher compression factors can be achieved as no postproduction editing will be made.

Even with the use of wide band communications it is not cost effective to transmit full bandwidth digitized video signals. The solution is to use compression to reduce the amount of data transmitted to more manageable amounts. These techniques are applied differently to the sound and video signals because the nature of the signals and more importantly the human response to them is not the same. As can be seen from Figure 1.1 the amount of data generated from digitized video signals is enormous, with approximately 1.3 Gbps streams for high definition television (HDTV), 270 Mbps for 4:2:2 video, and 135 Mbps for PAL quality video (4:2:0). Note that a typical high-speed digital telecommunication network operating at 155 Mbps (STM-1) would not have the capacity for even 1 digitized 4:2:2 video channel without compression!

It should be noted that all compression techniques introduce a variable degree of error to the original signal, with the reconstructed signal not being exactly the same as the original signal due to the introduction of irreversible quantization error during the encoding process. The purpose of the encoder is to minimize the amount of data to be transmitted

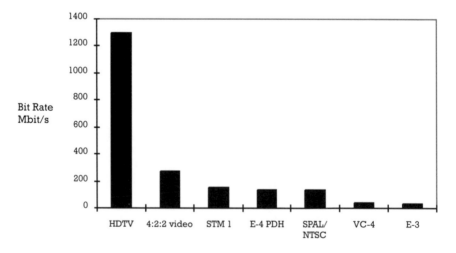

Figure 1.1 Uncompressed digitized video data rates.

(or stored) to a level consistent with the application. The decoder then has only to reconstruct this acceptable replica of the original signal from the quantized signal. The quantization schemes used in codecs are lossy in that they cause an irreversible loss of information due to quantization uncertainty or "quantization noise." However the degree of error can be tailored to user requirements, with a complete range from lossless transmission (and hence low compression) to lossy transmission (and high compression) possible. Quantization noise is more visible in the luminance component than the chrominance components, and also it has a larger effect on low frequency DCT (Discrete Cosine Transform) coefficients (the DCT is a mathematical function widely used in compression, as we will see in Chapter 4). Encoders can exploit these effects to minimize the error introduced into the signal while maximizing the level of compression. Depending on the application and the level of quality required different compression percentages could be achieved. In broadcasting applications, distribution and contribution circuits will in general require different compression schemes. This is because contribution circuits need to be of high quality to allow for editing prior to transmission, while final distribution circuits can tolerate more compression because there will be no further editing.

1.5.2 Constant and Variable Rate Sources

Pulse code modulation (PCM) is a very common digital modulation scheme used in modern communications. It is used in most audio encoders as the primary encoding process (see Section 4.13.4). In PCM a series of binary codes are generated which always approximate the amplitude of the signal sampled at that moment [2]. With a simple PCM digital system the overall bit rate is the product of the sampling rate and the number of bits allocated to each sample. Thus a constant bit rate results, whereas in digital television (and digital audio systems), the encoded information content has a variable bit rate that depends on the amount of picture detail. The difference between the encoded information rate and the total sampled bit rate is termed the redundancy. The purpose of compression in digital TV and audio is to remove the redundancy from information signals and in doing so reduce the bit rate required to digitally represent the information signal. However this lack of redundancy makes the compressed signal much more sensitive to any errors introduced in, for example, transmission or storage. The compressed signal requires the usage of powerful digital error correction techniques to protect the compressed signal from such errors and this process adds back in some redundancy, but this cannot be avoided if the compressed signal is to be protected from errors during transmission in noisy channels [3].

1.6   Intracoding and Intercoding Techniques

A digital video signal has four associated properties: it has a magnitude value for each sample, a time value, and a vertical and horizontal spatial value. Compression can be applied to any or all of these properties. However two main types of compression are used.

1.6.1   Intracoding

If a sequence of picture frames is compressed without reference to any other picture frame, the time value is not used as a reference parameter during compression, and compression is only applied to spatial values. This type of compression technique is then termed intracoding (within frames), and as we shall see later in Chapter 3 the JPEG standard for still pictures uses intracoding techniques to compress the data rate of information signals.

1.6.2   Intercoding

If the compression mechanism takes account of time and references some picture frames to others at a different time, a massive saving in bit rate can be made due to the temporal redundancy of video. This is termed as intercoding (between frames) and is used in the MPEG compression toolkit. It should be noted that editing of intracoded material is easier than that of intercoded material as the compression technique will not alter the content of frames in relation to time.

As was mentioned above, it is very important to match the amount of compression used to the actual process in hand, using as a decision tool the critical parameters of the signal. As can be seen in Figure 1.2 there are very different technical requirements between the acquisition of information (camera) and the secondary distribution (broadcasting) of the signal to the end consumer [4]. As a result MPEG-2 and other digital standards are used for these different purposes depending on the channel bandwidth available, the quality of image required, and the number of services to be delivered. We will explore the different compression standards in the next few chapters.

Process	Why use compression	Important Parameters
Acquisition	Digital acquisition and mobility	Image quality
Contribution	Interfacing to telecommunications networks	Image quality, low delay, cascading equipment
Editing	Increased productivity from networking	Ease of operation, image quality
Storage	Increased storage capacity	Quality, ease of retrieval
Primary distribution	Network cost and capacity, news service	Image quality, price per channel, network management
Secondary distribution (broadcasting)	Spectrum efficiency, new services, quality, new channels	Quality and quantity of services and channels, consistency in market

Figure 1.2 Broadcasting processes and compression requirements.

References

[1] de Bruin, R., and J. Smits, *Digital Video Broadcasting Technology, Standards, and Regulations*, Norwood, MA: Artech House, 1998.

[2] Kennedy, G., *Electronic Communication Systems*, New York: McGraw-Hill, 1985.

[3] Watkinson, J., *Compression in Video and Audio*, Focal Press, 1995.

[4] Neilsen, O., and N. Eriksen, *A Broadcasters Guide to MPEG*, RE, 1996.

CHAPTER

2

Contents

Digital Encoding of Television Signals

2.1 Objectives of Digital Compression

Compression, or bit rate reduction, is the term generally given to any process that reduces the amount of data associated with an information signal. It can be applied to video and audio signals, and is used to make a process more economical and practicable. In this chapter an introduction will be given to some of the more common standards used in the encoding of video and audio signals. It is the intention to leave the treatment of the MPEG standards until later chapters; such is the importance of these encoding standards.

It should be noted that image compression in television is not a preserve of digital television only. Lossy compression techniques are used in analog television systems. Interlace can be described as both a scanning and a compression technique. The reduction of bandwidth of the color difference signals of analog television standards such as PAL, NTSC, and SECAM can also be described as a form of compression. However it is only with the digitization of pictures that the enormous compression gains available today are possible. This is due to the algorithms and models that can be applied only to digital information signals to remove redundancy within the audio or video signal.

It is generally accepted that digital encoding standards that have as much commonality as possible between 525-line and 625-line television systems are desirable for reasons of equipment compatibility, economy, and source material interchange. Standards should be flexible enough to allow for possible advances in encoding and decoding technologies, and a move toward higher resolution receivers with 16:9 aspect ratio widescreen television. It was agreed by the International Telecommunication Union (ITU) radiocommunication assembly that a system based on the coding of component video would be the best standard to achieve these objectives. As a result Recommendation ITU-R BT.601 [1] was drafted based on the coding of the analog luminance and color difference signals—or if used, the red, green, and blue signals: R, G, B, which are generated from a studio camera (see Figure 2.1). Recommendation 601 therefore defines a set of encoding parameters for a digital component video system that can be used with both 525/60- and 625/50-television systems. These encoding parameters define the clock frequencies, the bandwidth of the component signals, the quantizing levels, the coding standard used, the position of the so-called active picture area, and the phase of the sample clock relative to reference syncs.

Any transmitted picture is sampled in time and also in the horizontal and vertical directions by a camera. The number of pixels (brightness

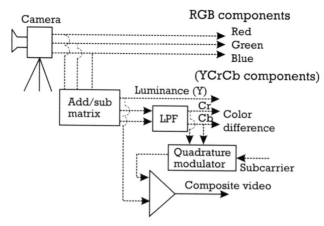

Figure 2.1   The main types of analog video.

changes, also termed picture elements) occurring along a horizontal line is dependent on the system horizontal resolution, while the vertical resolution is dependent on the number of lines used in the system. The final bit rate of a digital signal is dependent on the sample rate and hence sample rate reduction may be used to reduce the bit rate. However, as will be seen, it introduces irreversible loss of resolution, and thus the system designer must trade picture quality with bit rate in deciding the operating parameters of the encoder.

Some of the more common encoding/compression standards and recommendations will be summarized in this chapter.

2.2 Sampling Formats

In Chapter 1 reference was made to the bit rate that could be expected from the more common digital television formats including the commonly used 4:2:2 and 4:2:0 formats. These formats mentioned above, use numbers to indicate the ratios of the respective resolutions of the component signals. A brief description now follows of two of the more common component digital formats. Later in this and subsequent chapters reference will be made to 4:2:2 and 4:2:0 signals, as they are used in digital MPEG-2 encoders prior to transmission.

4:2:2 format: This is a video format widely used for interfacing and processing within studios. It is often incorrectly identified as CCIR Rec. 601, however that standard is more general and covers the whole family of formats at different sampling rates. Format 4:2:2 specifies the respective ratios of the resolution of the luminance and color difference signals. In this format the color difference signals are sampled horizontally at half the rate of the luminance component sampling, while the vertical sampling rate is the same for all components [2].

4:2:0 format: This format (as we will see later in this chapter) is widely used in digital television transmission systems, when bandwidth is limited. The ratio of the color difference (or chrominance) signals is reduced vertically to half that of the horizontal resolution by sampling the color difference signals only on every other line. This process reduces the amount of data generated from encoding and generates pictures that match more accurately the human eye response and the resolution of present day analog PAL I type receivers. The color difference signals are sampled vertically at only half the rate of the horizontal sampling of these signals. This results in a picture color resolution that is subjectively equal horizontally and vertically. It is an extremely important format and is used widely with many MPEG-2 encoding systems (as we will see in Chapter 4).

2.2.1 Rec. 601 (ITU-R BT.601-5)

This recommendation, as mentioned above, specifies methods for digitally encoding video signals using the luminance and two chrominance components (also known as color difference signals). It is based on using a standard 13.5 MHz sampling rate in the analog to digital converter for present day transmission systems [3], but allows for extension to higher resolution 16:9 aspect ratio using an 18 MHz sampling rate. To try and achieve as wide an application as possible it specifies that the number of active video samples per line be the same for 525- and 625-line systems. Also, whether 8-bit or 10-bit sampling is used, the 8 most significant bits are treated together firstly.

As can be seen from Figure 2.2 below, in 4:2:2 sampling of a video image, the light outputs of a camera consist of the three colors red, green,

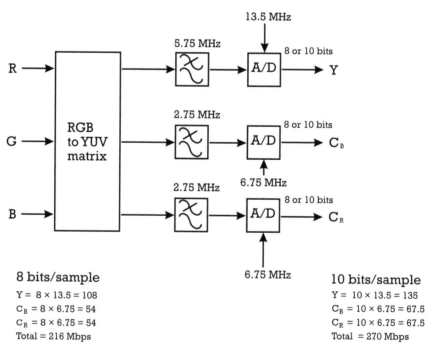

8 bits/sample
$Y = 8 \times 13.5 = 108$
$C_B = 8 \times 6.75 = 54$
$C_R = 8 \times 6.75 = 54$
Total = 216 Mbps

10 bits/sample
$Y = 10 \times 13.5 = 135$
$C_B = 10 \times 6.75 = 67.5$
$C_R = 10 \times 6.75 = 67.5$
Total = 270 Mbps

Figure 2.2   4:2:2 sampling.

and blue. These colors are first passed through a matrix to derive electrically the analog luminance (*Y*) and color difference signals (*Cr* and *Cb*). Following this stage they are low pass filtered, and then sampled at the specified 13.5 MHz rate for the luminance component and 6.75 MHz for the *Cr* and *Cb* components, to generate a digital representation of the original analog components. If 8-bit digital words are assigned to each sample a total bit rate of 216 Mbps results for the three components, while a 270 Mbps bit rate results at 10 bits per sample. This is termed "4:2:2 sampling." Compression systems used in transmission (such as those used for broadcasting DTV to the home) always reduce any 10-bit samples to 8-bit samples prior to the application of compression. This is because there is no need to assign 10-bit sample resolution to the quantization process, as it will be lost when displayed on a domestic television receiver (see Section 4.2).

2.2.2 Quantization Levels in 8-Bit Sampling

The allowed quantization states for luminance and color difference signals are shown in Figure 2.3, for 8-bit uniform quantization. When 8-bit uniform quantization is used there are 2^8 or 256 possible quantization levels for the signals to take. Levels 0 and 255 are reserved normally for synchronization, leaving levels 1–254 for video. In the luminance component signal level 16 is defined for black level, and level 235 for peak white level. To allow for signal overshoots and slight level-setting errors, margins are added so that the signal can occupy 220 levels only. The color difference signals are allowed (after quantization) to occupy 225 levels with the zero level defined as level 128. As we will see in Section 2.7, quantization error is added whenever a digital sample is used to represent a continuously varying analog signal. However this loss of resolution is not very significant, as it does not introduce any perceptual differences when monitored on a television receiver.

2.2.3 Redundancy and Compression

As mentioned in Section 1.3 and shown above in Figure 2.2, a 4:2:2 sampled signal generates an uncompressed 270 Mbps stream when sampled at 10 bits per sample, and similarly a 4:2:0 sampled stream generates 202 Mbps approximately. Clearly this data rate exceeds the capacity of common transmission channels as illustrated in Figure 1.1 and as a result to find an application, the data must be compressed to a lower bit rate. This can be achieved by exploiting the redundancy characteristics of video images.

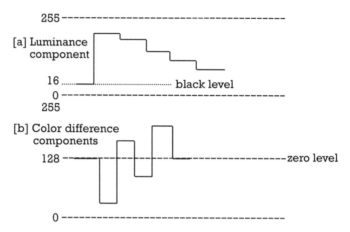

Figure 2.3   8-bit quantization levels for component signals.

Picture elements, or pixels, exhibit spatial redundancy within frames and temporal redundancy in successive frames. Individual pixel values are correlated with those of neighboring pixels and hence there is some measure of predictability of a pixel value given an adjacent pixel value.

All nonrandom digital signals display entropy redundancy whereby some values occur more often than others. Thus by assigning short codes to the more frequently occurring values, this feature can be exploited to achieve a coding gain. This feature will be seen to be exploited later in Section 4.6, by using run length coding and variable length coding to dramatically compress signals in the MPEG-2 compression algorithms.

The compression system chosen must also take advantage of psychovisual redundancy between the eye and brain. This allows the bit rate reduction system to exploit the human response to spatial resolution and temporal resolution, as the eye-brain combination has limits to the degree of finesse and tracking that it can assign to detail and movement, respectively.

2.2.4 Compression Efficiency

The highest levels of compression efficiency are achieved when component signals are used. The component signals may be in the common format of luminance and color difference or in the format of red, green, and blue. With television signals the commonly found format is the luminance and color difference signals which have been sampled according to the parameters defined in ITU-R Rec. 601.

If a signal has been encoded from component format into composite format (PAL, NTSC, or SECAM) then a very careful procedure must be used to decode the components from the analog composite signal prior to input into a digital encoder. This is because any residual subcarrier left in the luminance component from the composite signal will be interpreted by the digital encoder as high frequency moving detail. This moving detail will use significant amounts of valuable encoder bit rate to encode, in vain. This moving detail is an artifact that reduces the amount of data available to encode the rest of the information signal and degrades the quality of the compressed signal. The effect can be reduced by first pre-processing the composite analog signal prior to digital encoding. However, it is recommended where possible to encode component signals in preference to composite signals.

2.3   ITU-R Rec. 656 Interface

This recommendation defines an interface standard for digital video signals sampled according to the 4:2:2 definitions of Rec. 601. There are two possible techniques that can be used: bit parallel and bit serial. Rec. 656 has defined interfaces for the bit serial and also for the bit parallel connections. The serial interface requires that the data be packaged in a way that could also be used for the parallel interface.

The parallel interface is now rarely used to interconnect equipment but is still widely used inside equipment to interconnect, for example, circuit boards. However, serial digital component video is the current interconnect standard between equipment, for example, within a studio.

2.3.1   Serial Digital Component Video (4:2:2)

Serial digital component video (SMPTE 259M/EBU Tech. 3527) is a very important transmission format as it is widely used internally by broadcasting organizations. It is an uncompressed stream of digital video transmitted over a single wire and is of high quality. Often referred to as the "4:2:2" component, it is often the input to compression equipment prior to transmission and is used in digital studios. The term D-1 is a tape format only, but is widely referred to as the actual video format that the recorder uses. The actual video formats could be either parallel or serial.

Serial digital video transmits the video signal in component format with the luminance and two chrominance components separated. The reference 4:2:2 refers to the ratios in which the components are found.

As was seen in Figure 2.1, the digital luminance component (referred to as *Y*), and the two digital chrominance components (referred to as *Cr* and *Cb*) are electrically derived from the three full bandwidth analog color component signals, R, G, B (red, green, blue), which are available from cameras and other sources. As the ratio suggests, there are twice as many luminance or *luma* (Y) samples as each chrominance or *chroma* (Cr and Cb) samples. This is due to the fact that there is greater psycho-visual response to luminance information than chrominance information and hence the extra importance of this component. In fact for television type viewing only about 1/5 of the luminance bandwidth is needed to represent the chrominance information adequately for viewing. However for computer viewing applications the ratio is typically closer to 1:1.

The number of luminance samples at 720 per digital active line, is the same for both the 525-line, 60 field/s, and the 625-line, 50 field/s systems, refer to Figure 2.4. Similarly for both the 525- and 625-line systems each of the chrominance components have 360 samples per line.

However the two systems differ slightly in that the total number of samples per line is not the same in both systems, with 1,716 samples in the 525-line system, and 1728 samples in the 625-line system. This is due to the different number of digital blanking samples in both systems, with 276 samples in the 525-line system compared with 288 in the 625-line system. It should be noted that 4:2:2 sampling does not have equal chrominance resolution vertically and horizontally. The vertical resolution of chrominance and luminance is the same, while the horizontal resolution of chrominance is half that of luminance.

Parameter	525-line system	625-line system
No.of active video samples per line		
- Y signal	720	
- Cr signal	360	
- Cb signal	360	
Total No. of samples per line	1716	1728
Sampling frequency		
-Y signal	13.5 Mhz	
- Cr signal	6.75 MHz	
- Cb signal	6.75 MHz	

Figure 2.4   Number of samples in 4:2:2 sampling.

On each line there is a four sample code showing the End of Active Video has been reached (EAV), this is followed by the Horizontal blanking region and then another four sample code illustrating the Start of Active Video (SAV) which immediately precedes the 1440 luminance and chrominance samples.

The horizontal blanking interval can contain ancillary data or EDH data (error detection and handling) but usually contains alternatively 80 (hex) and 10 (hex) for 8-bit systems or 200 (hex) and 40 (hex) for 10-bit systems.

Each frame in 4:2:2 component video comprises lines of blanking, and active video lines as shown in Figure 2.5. Note that 525-line systems have only 486 lines of active video while 625-line systems have only 576 lines of active video.

2.4 4:2:0 Sampled Signals

The 4:2:0 sampling of signals is very important for two interconnected reasons. First, 4:2:0 is the color format specified by the main profile at

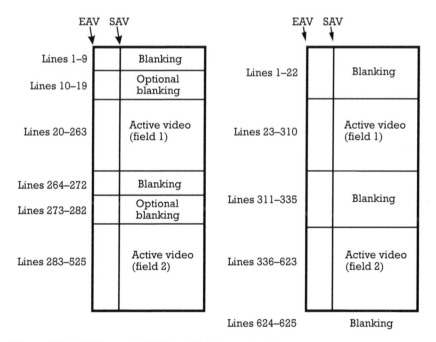

Figure 2.5 525-line and 625-line 4:2:2 frame structure.

main level (MP@ML) by MPEG-2. This is expected to be the most widely used profile within MPEG-2 for standard definition television (SDTV). Thus all other formats including 4:2:2 will have to be processed into this format prior to MPEG-2 encoding. Secondly, the resolution of 4:2:0 is approximately the same as that of standard delay line PAL receivers. It is optimized for use with standard PAL equipment, and so data is not needlessly encoded only to be lost in the receiver display.

With 4:2:0 sampling the vertical resolution of the chrominance is only half that of 4:2:2. This is achieved by reducing the chrominance sampling frequency by half. Note that with 4:2:0 sampling the chrominance resolution is subjectively equal vertically and horizontally.

2.5 ITU-Rec. 723

This standard has been developed for high quality digital television transmission over broadcast contribution and distribution systems. It is typically used internally by network operators on digital telecommunication networks such as fiberoptic and microwave link systems, and is not intended for broadcast to the consumer over terrestrial broadcast systems. It is a digital standard [4], designed according to ETSI specification ETS 300 174/ITU-R Rec. 723, and is commonly referred to as CCIR Rec. 723. It has the advantage of being fully compatible with 34/45 Mbps plesiochronous digital hierarchy (PDH) telecommunication networks.

It produces visually lossless compression results at bit rates between 34–45 Mbps, which corresponds to compression factors of between 5.5 and 4 to 1. The topic of lossless and lossy compression techniques is treated in more detail in Section 3.3.3, later. CCIR Rec. 723 can also carry multiple digital audio channels, teletext, and PALplus signals, as well as the main component and composite video formats. It has the advantage that the decoded signal can be easily edited and mixed. However multiple generation encoding and decoding is not recommended in post-production. It has proved to be very reliable in use and produces excellent quality images on all types of content material. It is however wasteful of bandwidth allocating 34–45 Mbps to 1 video service, and as a result high profile MPEG-2 distribution systems are beginning to replace it where bandwidth is a premium.

2.6 Encoding Signals for DTV Broadcasting

The main points to note from this chapter for application to the broadcasting of digital television to the consumer are that:

1. Compression techniques are not new, and not used only in digital television.

2. Digital compression removes the redundancy in an information signal using human perception models as well as purely technical compression techniques.

3. MPEG-2 is the preferred encoding standard for standard definition television.

4. 8-bit sampling will be used in the MPEG-2 encoder even if the source material has 10 bits per sample.

5. Composite signals can have artifacts present that reduce the efficiency of the encoder and therefore should be pre-processed prior to digital encoding.

6. Luminance and chrominance component formats are the preferred encoder input because with these signals high compression can be achieved with excellent quality of image.

2.7 Digital Audio Coding

Digital audio coding is performed in conjunction with any digital video coding scheme, such as the MPEG family toolkit of digital compression techniques. However digital audio can often be the main information source to be broadcast, as will be seen in Chapter 13, which deals with digital radio and in particular digital audio broadcasting (DAB). As with any digital process the more important concepts when dealing with audio signals are the sampling rates or frequency and the number of quantization levels.

This section will also deal with the internationally accepted digital audio interface standard for the serial transmission of linearly represented digital audio data. It is referred to as AES/EBU and is commonly found in studio equipment and as an input interface into broadcast equipment, such as DAB encoders.

2.7.1 Sampling and Quantization of Audio Signals

An audio signal is a constantly varying analog signal with amplitude values that are widely differing in value and have many frequency components. To represent such a signal with digital binary codes requires a process that can capture or sample the signal at discrete moments in time, and also be able to incorporate the dynamic range of the analog signal in amplitude for all of the frequency components.

The digitization of an analog audio signal employs time sampling of the signal and quantizing the signal into discrete amplitude values. This process is performed for the infinitely variable analog waveform and results in a series of discrete values in time and amplitude. Provided the analog signal is band limited to a finite frequency value by low pass filtering the signal, then it is possible to sample the signal at twice the highest frequency component without loss of information.

2.7.1.1 Quantization error

When quantizing an amplitude sample value of the analog signal into a digital word, there is a degree of resolution for each sampled value that is set by the word length of the digital code. (The word length may be thought of as simply the number of bits assigned to represent any digital sample value.) However the analog signal has a continuously varying amplitude value and when the digital quantizer sets the discrete levels that are assigned to the amplitude values there is the possibility of error. This happens whenever the analog signal amplitude falls between the discrete levels set by the quantizer, then the sample value at that instant will be rounded to the nearest discrete level of the quantizer. This approximation leads to a source of noise known as quantization error. It happens with all digitized signals, and can only be minimized to acceptable levels.

2.7.1.2 Digital Sample Word Length

The number of bits assigned to each digital sample will set the resolution of the quantizer and the accuracy of the digitization process. For example an 8-bit digital sample value would allow 2^8 quantizing levels or 256 discrete values, while a 16-bit sample value would allow 2^{16} quantizing levels or 65,536 discrete values. Clearly, the greater the number of bits available for a sample value the greater the resolution of the quantizer and the larger dynamic range of the audio signal which can be sampled.

2.7.1.3 Audio Signal-to-Noise Ratio

The signal-to-noise ratio (S/N) gives an indication of the available range of a particular word length. For an n-bit word, the signal to weighted noise ratio (in decibels) is given by the following relation,

$$S/N = (6n - 11) \quad \text{[dB]} \tag{2.1}$$

And therefore for a 16-bit system, the S/N ratio is equal to $((6 \times 16) - 11) = 85$ dB.

2.7.2 AES/EBU Digital Audio Interface

This digital audio interface was established by the Audio Engineering Society (AES) in collaboration with the European Broadcasting Union (EBU). It is a professional quality standard interface for the serial transmission of linearly represented audio. It is a sampling rate independent interface, and it can carry data from two independent audio channels with sample resolutions up to 24 bits. It is a commonly used interface to connect equipment within studios and is often used as the input signal into distribution encoding equipment such as stereo encoders for analog FM radio transmission and DAB encoders. Electrically it is a balanced signal of typically 2V – 7V in amplitude, with a characteristic impedance of 110 Ω (Ohms). It can be used in applications with up to approximately 100 meters of separation between equipment. It generally uses a three-pin XLR type audio connector.

There is a related digital audio interface known as SPDIF, which is an unbalanced electrical signal, with a typical amplitude of 0.5V, and is used in some consumer digital set top boxes (STBs). It is designed for connectivity over short distances of typically a few meters using phono connectors.

2.7.3 Common Sampling Frequencies

There are three commonly used sampling frequencies for digital audio encoding equipment. These sampling frequencies or sampling rates are used extensively in, for example, DAB encoders, refer to Chapter 13, and in MPEG audio encoders, refer to Section 4.13.4. The three most common sampling frequencies are 32 kHz, 44.1 kHz, and 48 kHz.

2.7.3.1 The 32 kHz Sampling Rate

This is the lowest of the three sampling rates adopted by the AES/EBU and it allows a practical bandwidth of 15 kHz for an audio signal, which is the minimum required for stereo TV sound and stereo FM radio broadcasting.

2.7.3.2   The 44.1 kHz Sampling Rate

This is the minimum required sampling rate to support a full 20 kHz bandwidth analog audio signal. This frequency, which is related to the line rate of television signals, is also used in compact discs.

2.7.3.3   The 48 kHz Sampling Rate

This is the highest sampling rate and that used by professional broadcast equipment. It can support the full 20 kHz bandwidth of analog audio signals. This sampling rate is related to the repetition rate of film pictures and will allow edited pictures to contain whole numbers of audio samples. This simplifies audio editing.

References

[1]   ITU-R BT.601, Studio Encoding Parameters of Digital Television for Standard 4:3 and Wide Screen 16:9 Aspect Ratios.

[2]   Steinberg, V., *Video Standards, Signals, Formats and Interfaces*, Snell & Wilcox, 1997.

[3]   Rumsey, F., and J. Watkinson, *The Digital Interface Handbook*, Focal Press, 1995.

[4]   ETS 300 174/ITU-R Rec. 723, Transmission of Component Coded Digital Television Signals for Contribution Quality Applications at the Third Hierarchical Level of CCITT Recommendation G.702.

Some Common Digital Television Standards

3.1 Introduction

In this chapter some of the more common digital television and audio standards will be summarized. These standards find application today in many consumer products and will continue to find applications within the Internet and home entertainment. Emphasis will be placed on the MPEG family of standards as these are expected to have the maximum penetration into consumer products and services. It is proposed to give a more detailed description of the MPEG-2 standard, later, in Chapter 4. This is because MPEG-2 is the standard developed for broadcasting applications and as a result it is used in all DTV transmissions. It is not intended to discuss the digital standards that relate to the modulation formats used in different transmission media at this stage (channel coding). The relevant standards for channel coding will be discussed in later chapters (see Chapters 6 and 7). This chapter focuses on the international standards for compression, decompression, processing, and coded representation of images and associated audio. Among the more important and common standards are the JPEG standard, and the MPEG family of standards including MPEG-1, MPEG-4, and MPEG-7.

3.2 Moving Pictures Expert Group (MPEG)

In 1987 the International Standards Organisation (ISO) and the International Electrotechnical Commission (IEC) formed a joint technical committee (JTC1) to co-operate on the standardization of information technology (IT). There are subcommittees (SC) within JTC1 with different areas of responsibility for the planning and coordination of various IT standards. One subcommittee SC 29 has responsibility for the "coding of audio, picture, multimedia, and hypermedia information." Under SC 29 three working groups were formed to develop international compression standards for still pictures (WG 1), coding of multimedia information (WG 12), and the Moving Pictures Expert Group (WG 11).

MPEG is therefore the working group number 11 of Subcommittee 29 of the joint ISO and IEC technical committee JTC1. This results in the commonly used reference ISO/IEC/JTC1/SC29/WG11 for MPEG standards. MPEG has various subgroupings working on different areas of the standards, for instance, audio, video, systems, tests, digital storage media, requirements, and applications.

3.3 Joint Photographic Expert Group (JPEG)

This is a compression standard that was originally designed for the storage and transfer of photographic images. The joint photographic expert group (JPEG) standard [1] was developed by a joint ISO/CCITT committee with the intention of specifying the compression and decompression of single frame images, and it was later extended to *motion JPEG* to include low bit rate video signals. It does not however have any facilities for the inclusion of audio with the still image, and is therefore unsuitable for broadcast material; however, it is widely used in editing, where audio is usually handled separately, anyway.

The distinction between still images and moving images is thin, as it is easy to treat a video sequence as a series of still images. By coding each frame in the series separately and then in turn decoding the series in the original order, the video sequence can be reproduced. This technique has the disadvantage however of ignoring the temporal redundancy in all video sequences. By not taking advantage of the inter-frame relationship of video sequences JPEG does not use any intercoding compression techniques (see Section 1.5) to remove this redundancy. This also makes the standard unsuitable for some broadcast applications that require high levels of compression to match the limited bandwidth available within the transmission channel.

3.3.1 JPEG Compression—An Overview

The type of compression applied in JPEG is intracoding, as it treats each picture frame, independently, frame by frame. This allows it to be used where editing is required, as each compressed image has no information about future or past images.

The need for data compression for even a single image becomes apparent when we consider the amount of data typically generated to represent a single picture image or frame. A 720 × 480 pixel (picture element) image stored using 24 bits per pixel (8 bits per component for the red, green, and blue color components) will require approximately 1.036 Mbytes of memory, without using compression.

3.3.2 Lossy and Lossless Compression

Two general types of data compression exist, lossy and lossless. Both are used to some extent by the JPEG (and also by the MPEG) "toolkit" of compression techniques. JPEG makes use of lossy data compression schemes to discard information and rely on human psycho-visual properties to ensure that the distortions produced in the image from compression are not noticed. As the name suggests lossy compression schemes inevitably add error to the decompressed signal, as information is lost in the encoding process.

JPEG also uses some well known lossless data compression techniques such as run length coding, Huffman coding, and predictive coding to reduce the amount of data generated by the encoding process. Lossless data compression schemes rely on reducing the redundant information in the data while representing the data with as few bits as possible. As the process is lossless no information is lost when the signal is decompressed [2].

The amount of data compression obtained using JPEG can be varied by changing the quantization levels (see Section 2.2.2) of the quantizer in the encoder. As we will see in Chapter 4, this changes the number of bits that are assigned to each portion of an image, and therefore the final amount of data required for this image. JPEG can achieve excellent quality images with compression ratios of the order of 10:1, and it is adequate for viewing at the quality level of a typical videocassette recorder with compression ratios of up to 32:1.

A JPEG compression unit is shown in Figure 3.1. A discrete cosine transform (DCT) unit transforms the blocks into coefficients that allow redundancy to be extracted (see Chapter 4 for a better description of the DCT, zigzag scanning, and variable/run length coding). A weighting unit next removes psycho-visual redundancy before a zigzag scanner places the coefficients in ascending spatial frequency order. This allows the identification of common bit codes for compression in a variable length and run length coder. The compression factor required establishes the quantization step size. Finally the data are assembled into packets and assigned with headers for transmission. The reverse process is shown in a JPEG decoder in Figure 3.2.

3.4   MPEG-1

MPEG-1 refers to the ISO/IEC 11172 International Standard [3]. MPEG-1 was the first standard produced by the working group and it is intended for applications such as computer images and graphics. It is actually

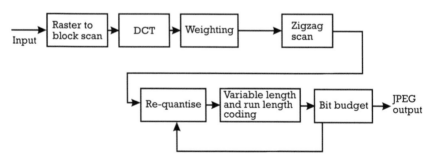

Figure 3.1   A JPEG encoder.

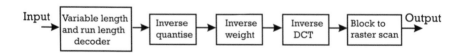

Figure 3.2   A JPEG decoder.

slightly lower in quality for movies than a videocassette, but being digital, it suffers less from degradations. It is suited to compact disc applications on computers and supports coding bit rates up to approximately 1.5 Mbps. MPEG-1 is a derivative of the H.261 videophone specification and is therefore best suited to simple source materials such as computer graphics and low motion pictures such as talking heads. It does not support interlaced video and has a limited degree of resolution; however, it is useful for video servers and CD-I (compact disc interactive) type applications. It has features such as random access, robustness, and ease of editing of video, however, that make it suitable for the storing and playback of digital video. It is popular in the computer games market and is used for movies stored on CD-ROM (compact disc read only memory). The accompanying digital audio signal is produced in stereo and at a bit rate of 192 Kbps. To achieve the required bit rate of approximately 1.5 Mbps the encoder must eliminate redundancy in the source material and also a large amount of the signal entropy (see Chapter 2). This is achieved, as we will see, by subsampling the source material and discarding alternate fields. These actions result in a lossy compression system that is not suitable for high quality applications.

3.4.1 MPEG-1 Compression—An Overview

The MPEG-1 standard defines generic audio and video source coding with a large amount of flexibility that supports a large range of coding parameters. The standard specifies the syntax and semantics of the bitstream that is produced by the encoder while not specifying the actual encoding process. It also specifies the signal processing in the decoder.

The standard also supports video only applications and audio only applications. Digital Audio Broadcasting (DAB) is a popular example of layer II of the standard (see Section 4.13), which applies only to audio at present. (However, it should be noted that DAB has been proposed by some radio broadcasters as a means to deliver JPEG encoded pictures and even MPEG-1 encoded video sequences, to support the audio content of radio broadcasts.)

The MPEG-1 standard is broken into five parts: systems, video, audio, compliance testing, and software testing. The system specification defines everything needed for multiplexing and synchronizing the video and audio bitstreams. This provides the mechanism for storage and transmission of the bitstream but it must only contain one audio and one video program in the bitstream. The video specification defines the syntax of the encoded video bitstream and its semantics. The audio specification defines the syntax of the encoded audio bitstream and its semantics. The audio specification is divided into three layers of complexity which are suited to different applications, from those involving mild compression to a more complex layer that supports DAB and uses more compression, to a layer employing the most compression and suited to telecommunication applications (see Chapter 4 for more details). The compliance specification defines how to interpret compliance for the earlier three parts, and provides guidelines on how to determine compliance in bitstreams and decoders (but not encoders). Part 5 of the specification provides a software simulation example of an encoder and decoder for audio and video.

In order to achieve high compression ratios the MPEG-1 encoder must pre-process source material from, for example, a (CCIR-601) 4:2:2 signal to a 2:1:0 signal by first subsampling the signal and by then disregarding alternate fields. It has by then produced what is termed as source input format (SIF), and this is now suitable for MPEG-1 encoding [4]. The process of discarding alternate fields and subsampling of the chroma signals and luminance signal reduces the data rate to approximately 3/16 of the original data rate. An opposite postprocessor will be used to reproduce a 4:2:2 signal in an MPEG-1 decoder. As this preprocess has reduced the signal bit rate enormously, before using the standard MPEG toolkit for encoding there will be an unavoidable loss in signal quality as information has already been thrown away to achieve compression. This makes the format unsuitable for action and sport and for distribution of broadcast quality signals. However the reduction in bit rate from a 4:2:2 signal at 168 Mbps (assuming 8-bit coding) down to a 1.5 Mbps data rate is the main attraction of MPEG-1 coding.

There is a hierarchical structure to the MPEG family of standards, and as a result any MPEG-2 decoder will be able to decode any MPEG-1 signal. However, an MPEG-1 decoder will not be able to decode an MPEG-2 signal [5].

3.5 Frame-based and Object-based Coding Standards

MPEG-1 and MPEG-2 are known as frame-based standards, however the trend within MPEG is to develop other standards which have a more general and powerful way of coding audio-visual content, in which data is described in terms of *objects*. Frame-based coding and description schemes provide limited capabilities in terms of the access, manipulation, and identification of individual objects. Object-based coding and description offer the user a new range of capabilities where objects are separately coded and described, giving a new dimension to playing with, creating, and accessing video content. This has enabled the development of richer applications in the area of content-based database retrieval, multimedia broadcasting, games, remote surveillance, and advanced personal communications. Due to industry demand and the availability of adequate technology, MPEG is currently developing two new and important object-based standards: MPEG-4 and MPEG-7. These standards are not for application to real video from a camera, except as part of an assembled screen containing other objects.

3.6 MPEG-4

MPEG-4 is being developed as a solution for audio and video media scene composition and transport. It also will be able to manage and protect content. It is believed that it will be an important standard in DTV for handling data.

The MPEG-4 standard uses an object-based composition approach, supporting two-dimensional (2D), arbitrarily shaped, natural video objects as well as synthetic data. Synthetic data includes text, generic 2D/3D graphics, and animated faces, enabling content-based interaction and manipulation. The integration of the objects is taken care of by the MPEG-4 powerful systems layer, which also provides much desired interactive capabilities.

MPEG-4 is an ISO/IEC standard (designated ISO/IEC 14496) being developed by MPEG (Moving Picture Experts Group). MPEG-4 has built on the proven success of three fields: digital television, interactive graphics applications (synthetic content), and interactive multimedia (World Wide Web, distribution of and access to content). It is planned that MPEG-4 will provide the standardized technological elements enabling the integration of the production, distribution, and content access of these three fields.

MPEG-4 provides standardized ways to represent units of aural, visual or audio-visual content, called media objects. These media objects can be of natural or synthetic origin; this means they could be recorded with a camera or microphone, or generated with a computer. It describes the composition of these objects to create compound media objects that form audio-visual scenes. MPEG-4 can multiplex and synchronize the data associated with media objects, so that they can be transported over network channels providing a QoS appropriate for the nature of the specific media objects, and it describes ways to interact with the audio-visual scene generated at the receiver's end.

3.7   MPEG-7: A Multimedia Content Description Interface

Audio-visual information is becoming available more and more easily from sources around the world, whether on the Internet, through Digital Audio Broadcasting (DAB) transmissions, Digital Television (DTV) or from CD-ROM. More and more people want to search for and use audio-visual content. However before this information can be used it must be located. As more information becomes available this search becomes more difficult and takes longer to perform. This has led to the development of a scheme to search quickly and efficiently for multimedia material by using standardized descriptions of various types of multimedia material.

MPEG-7 will specify standard multimedia description schemes and descriptors, as well as a description definition language to extend the range of description schemes, with the target of solving the problem of efficiently identifying relevant audio-visual information.

The description will be associated with the data itself, to allow fast and efficient searching for the material that is requested by the user. The standard does not comprise the automatic extraction of the content descriptions. It does not specify the type of software program which can make use of the description, thus allowing Internet search engines to be written in the future with more efficient algorithms to search for content at higher speeds [6].

Although MPEG-7 is still in a requirements stage, many companies, ranging from broadcasters and database owners to telecom operators have already shown a large interest. It is expected to have fundamental effects on archiving material, and will change the way information is located within, for example, libraries and news departments.

3.7.1   Metadata

Metadata is a term which is attributed to Nicholas Negraponte and appears in his book, *Being Digital*. Metadata can be defined as data describing other data. It is currently the topic of much interest, and is used within search engines on the Internet to search Web sites [7]. Metadata can be associated with any information but is especially useful with audio and video content. Searching through metadata is much faster than searching through actual video and audio content. It is also important to realize that there are many possible classes of metadata, with some classes being much more important than other classes. For example the digital video flags described in Figure 2.5, that is EAV and SAV are very important to allow the display of digital video. Similarly the MPEG SI (service information) is also very important. Some types of metadata change often and must be repeated at regular intervals in a carousel, while other metadata is static. A detailed discussion on metadata is beyond the scope of this book, but is related to advances within the MPEG object-based coding technologies mentioned above.

References

[1]   ISO Joint Photographic Experts Group Standard JPEG-8-R8.

[2]   Woodward, W., "An Overview of the JPEG and MPEG Video Compression Specifications," *NCTA Technical Papers*, 1991, pp.135–141.

[3]   ISO/IEC JTC1/SC29/WG11 MPEG, International Standard ISO 11172,*Coding of moving pictures and associated audio for digital storage media up to 1.5 Mbits/s (1992)*.

[4]   Watkinson, J., *Compression in Video & Audio*, Focal Press, 1995.

[5]   Watkinson, J., *The Engineers Guide to Compression*, Snell & Wilcox, 1996.

[6]   "MPEG Update," *Broadcast Engineering*, April 1999, p. 20.

[7]   Gilmer, B., "The EBU/SMPTE Task Force—Part IV, Wrappers and Metadata," *Broadcast Engineering*, April 1999, pp. 58–62.

Contents

MPEG-2: The Broadcasters' Choice

4.1   Introduction

The Moving Pictures Expert Group (MPEG) has developed the MPEG-2 standard, within Working Group 11 (of SC29 of JTC1) of the International Standards Organization. It is intended for application to the encoding of high quality video images and audio signals, with bit rates in the range of approximately 4–100 Mbps. It took as a basis both the ITU-T standard for videoconferencing and videotelephony (H261), and also the JPEG standard (see Chapter 3).

MPEG-2 is a frame-based video encoding standard as opposed to the MPEG-4 and MPEG-7 standards, which are object-based standards (see Section 3.4), and unlike the MPEG-1 standard it can support interlaced video. It can also support variable aspect ratios, which means that it is possible to transmit a pan-scan window, so that for example a 4:3 display can be selected by a decoder from a transmitted widescreen 16:9 picture [1]. MPEG-2 is a scalable set of tools for compression, and the scalability means that the signal can be decoded to various levels to enhance various features such as resolution or signal-to-noise ratio. These are among the reasons why MPEG-2 has been accepted worldwide by broadcasters as the digital encoding standard that will be the foundation of digital transmission systems into the new millennium.

The standard specifies only the data stream syntax and the decoding process, not the actual encoding process. This allows for a wide range of coding techniques to be used and therefore possible improvements in bit rate reduction (compression) technologies in the future. The only requirement for a coder is that it must produce a valid data stream. This means that future improvements in encoding technology should not result in decoding equipment becoming obsolete. As the decoding part of the standard is now defined and as long as manufacturers follow the MPEG-2 standard, any decoder will be able to benefit from future encoding improvements and possibly have better quality pictures with the same decoders

This chapter will also discuss the MPEG approach to audio encoding for digital television (DTV) and digital audio broadcasting (DAB). The different layers within MPEG-1 and MPEG-2 audio standards will be examined. Musicam, psychoacoustics, and compression techniques will be dealt with toward the end of the chapter. DAB will be discussed using these topics in a later chapter.

4.2   4:2:0 Sampling

In Chapter 2 we saw how encoding a 625-line component picture according to Rec. 601 could result in a bit-stream of 216 Mbps with 8 bits per sample, or even 270 Mbps with 10 bits per sample, termed 4:2:2 sampling. As was discussed in Chapter 1, this is not a realistic bit rate for transmission purposes as very few transmission media have this data capacity, and hence compression is needed. Sample rate reduction can be used to reduce the bit rate generated from the encoding process but it will affect the resolution of the picture.

For broadcast applications however, sufficient chrominance resolution can be provided even with half the sampling rate on the vertical chrominance component, resulting in what's termed 4:2:0 sampling. This sampling method results in a picture image of the same resolution as that of a standard analog PAL decoder using a delay line [2]. This method of sampling is generally accepted as the most applicable level and profile for

MPEG-2, and is termed Main Profile at Main Level or, MP@ML. For this reason the widely used 4:2:2 format must be converted to 4:2:0 format by vertical filtration prior to MPEG-2 encoding. Processing of the material into 4:2:0 format is performed to ensure that there is equal subjective chrominance resolution vertically and horizontally in accordance with the color perception of human vision. Some of the other important MPEG-2 toolkit of compression techniques will be discussed in the next sections.

4.3 Discrete Cosine Transform (DCT)

This is an orthogonal mathematical transform that is used to remove spatial redundancy in the sampled signal components, by concentration of signal energy into only a few coefficients. The DCT and it's inverse are easily implemented with digital signal processing (DSP) technology. In Figure 4.1 are shown a number of 8 × 8 blocks, containing 8 pixels and 8 lines from a 625-line 4:2:0 sampled picture. The complete image corresponds to approximately 704 pixels by 576 vertical lines, which results in 88 blocks horizontally and 72 blocks vertically. These 6,336 two-dimensional (8 × 8) blocks are each input into a discrete cosine transformation that maps the sampled values onto corresponding values in the frequency domain.

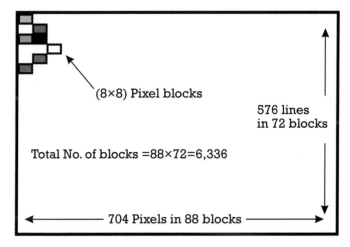

Figure 4.1 Digitized 625-line picture.

An example of the DCT processing of an 8 × 8 luminance block is shown in Figure 4.2, with an image shown of dark grey steps (top left) gradually fading to white (bottom right). Superimposed on the image are the corresponding 64 approximated digital sample values, generated according to quantizing and sampling rules referred to in Section 2.2. Note that according to these rules white corresponds to a sample value of approximately 235 and dark grey approximately 44.

One of the first steps in the DCT process involves finding the average luminance value of the samples and subtracting this value (in this example, 128) from each of the 64 samples to eliminate the average or DC component. This results in a new matrix of 64 integer values, some of which are negative in value.

Next the 8 × 8 matrix is applied to the DCT where another matrix of numbers emerges. This time the values that are produced are typically reduced in magnitude, but contain a number of extra decimal places. It should therefore be noted that the DCT process often results in more data being generated than was present in the 64 initial sample values, and hence on its own, the DCT often does not result in compression! However as we will see next, in combination with a quantization law and table, it can have a very high compression gain.

4.4   Coefficient Quantization

After the DCT the new 8 × 8 matrix contains coefficients that are often negative and close to zero in value. These values are then compared to a

44	53	62	71	79	88	97	106
53	63	74	85	95	106	117	127
62	74	87	99	111	126	137	149
71	85	99	113	127	142	158	172
79	95	113	127	143	159	175	191
88	106	124	142	159	177	195	213
99	117	136	156	175	195	215	234
106	127	149	172	191	213	234	255

Figure 4.2   Grey to white steps.

set of values in a quantization table (another 8 × 8 matrix) in accordance with a particular and fixed quantization rule. The purpose of this process is to reduce most of the DCT coefficients to zero if possible. Different quantization tables can be used by the encoder at any time to reduce or increase the amount of data that is being generated. This results in a non-fixed bit rate signal emerging, and we will see later how this is handled by a buffer. A typical quantization law (4.1) would be a linear law, which might for example round off the ratio of each DCT element and corresponding quantization element to an integer value.

$$F(i, j) = INT[DCT(i, j) / Q(i, j) + 0.5] \qquad (4.1)$$

Here $DCT(i, j)$ is the discrete cosine transformation value for element (i, j) of the 8 × 8 block, $Q(i, j)$ is the corresponding element from the quantization look up table, INT is the integer rounding function, and $F(i, j)$ is the resulting element after the process.

A typical example of the resulting reduction in elements by this process is shown in Figure 4.3. As can be seen most elements are now zero except for those elements in the top left-hand corner of the matrix, and as predicted these correspond to the lowest frequency elements of the original signal.

4.5 Zig-Zag Coefficient Scanning

The next step in the encoding process is to use a scanning technique to reduce the amount of data that represents the 8 × 8 matrix shown in

0	−25	0	−2	0	0	0	0
−23	6	0	0	0	0	0	0
0	0	0	0	0	0	0	0
−2	0	0	0	0	0	0	0
0	0	0	0	0	0	0	0
0	0	0	0	0	0	0	0
0	0	0	0	0	0	0	0
0	0	0	0	0	0	0	0

Figure 4.3 Reduction of DCT coefficients with quantization.

Figure 4.3. At this stage after the DCT, most coefficient values are now zero with only a few nonzero elements. The two dimensional image is now scanned in a zigzag fashion to produce a long string of coefficients, most of which are zero. Two different scanning methods can be selected, depending on the content material, and the coder can signal the method chosen to the decoder automatically [3]. A common type is shown in Figure 4.4, where a symmetrical search of horizontal and vertical coefficients is made.

The coefficient representing the lowest frequency or DC component is situated at the top left-hand corner of the matrix while the highest frequency coefficient is situated at the bottom right-hand corner. Typically there are many zeros in the lower right part of the matrix, and so the scan starts at the more important DC coefficient and ends at the less important lower right-hand corner following the predefined zig-zag scan.

4.6 Variable Length Coding and Run Length Coding

By counting the number of zeros (termed *run length coding* or RLC) between nonzero elements in this serial string and by applying a variable length code (VLC) to this number combined with the nonzero value preceding it, enormous coding gains can be made. The variable length code has a bias toward commonly occurring combinations of zeros and the preceding nonzero value, and allocates shorter code values to these

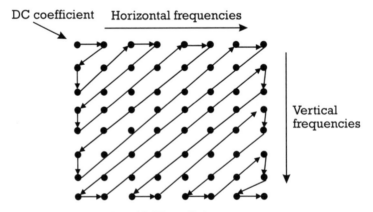

Figure 4.4   Zigzag scanning of DCT coefficients.

combinations. The two variable length codes most often used are similar to Huffman codes, and as such no confusion arises about where a variable length code ends and another starts.

4.7   Buffers

As mentioned above in Section 4.4, at different times the MPEG-2 encoder can produce a different bit rate signal, depending on the actual signal content and the amount of motion within the image. However, a constant data rate is required for transmission systems and hence an electronic buffer is used to smooth out fluctuations in data rate. This is achieved because the buffer can store data as it emerges from the encoder and output it at a constant data rate. This in the short term will smooth out any potential fluctuations in data rate due to changes in the content being encoded. The only risk is that the buffer may fill up and data may overflow if the amount of data input to the buffer cannot be controlled. However, this alone is not enough to control the amount of data being generated, and as mentioned earlier, the encoder can choose from more coarse (or less coarse) quantization tables to control the number of bits generated for each DCT coefficient. By combining the smoothing effect of the buffer with the variable data rate produced by different quantization tables, the encoder can effectively control the actual data rate of the signal as the information rate (content) of the signal changes from instant to instant.

4.8   Interframe Redundancy Compression Techniques

In the introduction to this chapter it was again noted that MPEG-2 is a frame-based encoding system, and it has some tools available to it to exploit the temporal redundancy of video signals from frame to frame. On a frame to frame basis pictures often change very little and MPEG-2 uses this fact to try and predict a frame from the value of another frame termed a reference frame, and this process is known as interframe prediction. The simplest prediction to make is to try and predict the values of a block in the exact same spatial position as those contained in the reference frame.

By using a differential pulse code modulation coder (DPCM) it is possible to quantize and transmit only the differences between the input and a prediction [2]. This prediction must be based on locally decoded output because the prediction must be repeatable in a remote decoder where the source pictures will not be available. Hence the coder will always contain a local decoder, which reconstructs the picture in the same manner as a remote decoder would do. The MPEG-2 coder subtracts the prediction from the input to generate a prediction error picture, which is transformed by the DCT, the coefficients quantized, and then coded as described earlier. As mentioned above the simplest prediction is the co-sited prediction of a block given a reference block, which is easily accomplished with a simple delay, but this method is only effective in pictures where there is no image movement within that block from frame to frame.

4.9 Motion-Compensated Interframe Prediction

This form of prediction is more sophisticated than the interframe prediction method just described and is used to offset any translational motion that may have occurred between the block being coded and the reference frame. It uses a shifted block from the reference frame as the prediction. As mentioned earlier the MPEG-2 standard defines only the decoding process and as a result different methods and algorithms can be used to perform the motion estimation. Quite considerable performance differences can be achieved between different manufacturer implementations. The most important parameter is the search area used to cover motion from frame to frame.

In Section 4.3 the basic coding unit for removal of spatial redundancy was defined to be an 8 × 8 block, however with MPEG-2 motion-compensation is usually based on a 16 × 16 block, termed a macroblock. This size is a trade-off between the requirement to have a large macroblock size in order to minimize the bit rate needed to transmit the motion representation or motion vectors, and the requirement to have a small macroblock in order to be able to vary the prediction process within the picture content and motion.

There are many methods available to generate a motion compensated prediction. These include forward prediction, where a macroblock is predicted from a past block, backward prediction, where a block is predicted from a future block, and intracoding where no prediction is made from the macroblock. These prediction modes are applied to MPEG-2 pictures depending on the picture type.

4.10 MPEG-2 Picture Types

In general, the term picture covers a coded entity. A picture can be either a frame or a field. It is possible to change dynamically between frame coding and field coding from frame to frame. Frame coding is preferred when a lot of detail, but little motion is present, and field coding is preferred when a lot of movement is present. There are three commonly defined picture types in MPEG-2 and these establish which prediction mode is suitable to code each macroblock of the picture.

4.10.1 I-pictures

I-pictures, or intra pictures, are coded without reference to any other pictures. They provide access points in the data bit stream where decoding can begin without reference to previous pictures. Spatial redundancy can be removed from the picture but temporal redundancy cannot, and hence limited compression gain can be achieved with these picture types.

4.10.2 P-pictures

P-pictures, or predictive pictures, can be coded using motion compensation and predicting from past I- or P-pictures. They may be used as a reference for subsequent prediction. With both temporal and spatial redundancy removal, P-pictures offer increased compression gain over I-pictures.

4.10.3 B-pictures

B-pictures, or bidirectionally-predictive pictures, use past and future I- or P-pictures for motion compensation. In order to achieve backward prediction from a future frame the encoder must re-order the sequence of pictures from the natural temporal picture order into transmission or bit stream order. This has the effect that the B-picture is transmitted after the future pictures which it references. This will introduce a delay depending on the number of consecutive B-pictures that are generated. It has the highest possible compression gain of all three picture types.

4.11   Group of Pictures (GOP)

The above-mentioned three types of pictures occur in a regular and repeating structure, which is termed a group of pictures. There are two parameters used to describe a GOP: N, the number of pictures in the group, and M, which is the spacing of the predictive pictures. A GOP is defined as a picture sequence that can be coded as an entity, and editing or splicing of the encoded data stream has to take place between GOPs.

4.12   MPEG-2 Profiles and Levels

The MPEG-2 standard is capable of many implementations, as the standard has been developed to allow this flexibility. However there are defined profiles that are a subset of the overall system, and are defined to allow the user tailor application requirements to encoding parameters. Some of the profiles or techniques are listed below [4] and the interrelationships between encoding parameters for different profiles and levels are shown in Table 4.1, together with the maximum total bit rate. By selecting a particular profile certain picture types may or may not be allowed.

Main profile allows all types of pictures and prediction. It is the most common profile used. It has a delay due to these pictures.

Table 4.1
MPEG-2 Profiles and Levels

Levels	Simple profile (SP)	Main profile (MP)	SNR scalable profile (SNRP)	Spatially scalable profile (SSP)	High profile (HP)
High (HL)		MP@HL 80 Mbps			HP@HL 100 Mbps
High-1440 (H14L)		MP@H14L 60 Mbps		SSP@H14L 60 Mbps	HP@H14L 80 Mbps
Main (ML)	SP@ML 15 Mbps	MP@ML 15 Mbps	SNRP@ML 15 Mbps		HP@ML 20 Mbps
Low (LL)		MP@LL 4 Mbps	SNRP@LL 4 Mbps		

Simple profile does not allow B-frames, so there is no backward prediction. This makes this profile a very low delay profile, as there is no re-ordering of pictures. It allows minimal complexity at encoder and decoder.

Spatial profile supports enhancement layers that carry the image at different resolutions using the spatial scalability tool. It allows the reproduction of the image with different resolution pictures depending on the decoder display.

High profile supports 4:2:2 sampled video, and as a result is seen as a migratory path by broadcasters from ETSI or CCIR Rec. 723 encoding (see Section 2.5) toward MPEG-2 for contribution and distribution circuits.

SNR (or S/N) profile adds support for enhancement layers of DCT refinement, using signal-to-noise ratio (SNR) scalability. It is a technique where two separate data streams are used to carry one video signal. If both streams are decoded together a high quality signal will be received, if one only is received the result is a picture with the same resolution but with a reduced quality due to reduced signal-to-noise ratio. It is believed by some that this may find use in terrestrial transmission to fixed aerial and portable aerial receivers. This would possibly enable high quality fixed or rooftop aerial reception, while simultaneously allowing portable reception at a reduced quality level, which would be better than no picture reception at all.

As can be seen above, a profile defines the amount of functions and compression processes involved. Within each profile a level is then defined which places constraints on parameters within that profile, the most common constraint being sampling rate. The levels differ primarily in the resolution available and the bit rate required. The main level (ML) is appropriate for standard definition television (SDTV) and most current interest is in the main profile at main level (MP@ML). Note that MPEG-2 decoders within a given profile and level are expected to be able to receive and decode inputs from lower profiles and levels.

4.13 MPEG Audio Coding

This section details and explains some of the MPEG family of audio encoding and decoding standards. It gives an insight into the reasons behind the development of the various standards that are used in different broadcasting applications. It then describes the DVB guidelines for using MPEG audio coding. It should be noted that both the MPEG-1 audio and the MPEG-2 audio standards could be used with MPEG-2 encoded transport streams for DVB broadcasting. As we shall see the main difference between the two standards is that MPEG-2 audio can carry more than two audio channels. The MPEG-2 audio standard is however backward compatible with all MPEG-1 audio decoding equipment. This means that an MPEG-1 audio decoder can decode the basic stereo information of a multichannel MPEG-2 audio encoded signal. It is important to note that there are three different layers associated with both of the current MPEG standards, and that layer II is the most applicable for broadcast applications.

Without using data rate reduction techniques, a digital stereo audio signal will typically generate a data rate of 1.4 Mbps. This is using standard 16-bit samples and an audio bandwidth of 44.1 kHz. With MPEG audio-coding techniques large compression factors can be achieved (up to around 12) without a perceived loss of signal quality.

MPEG audio basically describes the coding of audio signals using perceptual coding schemes. These techniques exploit the psycho-acoustic nature of the human ear response, to compress audio signals. In contrast, a dedicated speech codec exploits the properties of the human vocal tract, trying to maintain the intelligibility of the voice. These speech codecs have very low delays in processing conversations as required in telephony circuits. They also outperform MPEG codecs in compression of speech signals at very low bit rates. However, the MPEG codecs are suitable for high quality audio signals, including broadcast signals, and can achieve high compression factors while maintaining a superior quality to speech codecs.

4.13.1 MPEG-1 Audio

The audio subgroup of MPEG had the responsibility of developing a standard for the generic coding of audio signals that would be compatible with the relevant MPEG video encoding standards. The result of the work was a three layer standard for MPEG-1 audio, with different complexity levels for different applications, and is designated ISO/IEC 11172-3 [5].

4.13.1.1 MUSICAM (Layer II)

After intensive testing the ITU recommended using MPEG-1 layer for contribution, distribution, and emission, which are of course typical broadcasting applications. The sampling rates used in this standard are 32, 44.1, and 48 kHz. The corresponding bit rates are 32 to 192 Kpbs for mono audio channels and 64 to 384 Kpbs per stereo audio channel. This

MPEG-1 layer II standard is identical to the well known MUSICAM (Masking pattern adapted universal sub-band integrated coding and multiplexing) audio coding system [6], and the layer I system is best understood as a simplified version of the MUSICAM system. The MUSICAM system is used in digital audio broadcasting (DAB). (See Chapter 13 for more information.)

The MPEG-1 layer II system supports up to dual channel audio only, and uses relatively high sampling rates. In contrast, MPEG-2 layer was developed to extend the number of channels to 5 for surround sound capability, and by using lower sampling rates (1/2 the sampling rate of MPEG-1 audio) it can be used more efficiently on low bit rate applications such as speech signals.

4.13.2 MPEG-2 Audio

The MPEG-2 audio standard was developed to meet two main objectives. The primary objective was to extend high quality coding from two to five channels in a backward compatible manner. The MPEG-2 audio standard was adopted in 1994 [7] and provides for five high quality full bandwidth channels plus a narrow bandwidth low frequency enhancement channel. It has multichannel capability for applications such as multilingual broadcasting. It has surround sound capability with the loudspeaker positions, left, center, right, and left and right surround. This is often referred to as "3/2 stereo," as it contains 3 front and 2 surround channels.

The secondary objective of the MPEG-2 audio standard was to extend the capability of MPEG-1 audio to lower sampling rates. The purpose of this was to improve the quality of low bit rate signals such as speech signals at bit rates below 64 Kpbs.

4.13.3 Different ISO MPEG layers

As mentioned above the MPEG standards have three layers of complexity and performance for different applications. The ISO standards only define the format of the encoded data stream and the decoding process. This allows manufacturers the freedom to design improved psycho-acoustic models and audio encoders in the future, in the knowledge that MPEG decoders will benefit from the innovations yet to be designed.

4.13.3.1 MPEG Layer

MPEG, or ISO layer I, is also known as simplified MUSICAM, and is appropriate for mild compression applications at low cost. It is similar to NICAM (near instantaneous companding audio multiplex) digital stereo in that samples in each sub-band block are block compressed in accordance to the peak value of that particular block. It has the lowest encoding delay of the three MPEG layers. It also has the lowest performance, that is, sound quality per bit rate.

4.13.3.2 MPEG Layer II
The MPEG ISO layer is identical to the MUSICAM system, and is also used in the Digital Audio Broadcasting (DAB) radio transmission system (see Chapter 13). It is a generic coding standard for multichannel sound for digital video broadcasting and computer multimedia. It is the most important layer for broadcasting applications. Generally layer II encoders offer a higher sound quality at lower data rates than layer I encoders. This is achieved by using a more accurate Fast Fourier Transform (FFT) process and therefore a more accurate perceptual modelling process, and also by grouping scale factors to obtain maximum reduction in overhead bits. It has a longer encoding delay than a layer I encoder, but has a better performance.

4.13.3.3 MPEG Layer III
The MPEG ISO layer III is the most complex layer of the standard and is reserved for use with applications requiring either the highest quality reproduction of audio signal or the highest compression factor. Some telecommunication circuits use it at low bit rates, and it can be also used to generate higher data rates for broadcast applications. It is regarded as the most powerful layer of the MPEG audio coding techniques, as it gives the highest quality signal at a given bit rate, or the lowest bit rate signal for a fixed quality level. It is not yet in general use for broadcast applications, but it is in use with Internet based audio compression algorithms. It has the longest delay in encoding source material. This compression system has been suggested for the digital radio mondiale consortium broadcast system to be introduced for AM radio (see Section 13.12).

4.13.3.4 MPEG-2 AAC
The latest MPEG audio development is the MPEG-2 Advanced Audio Coding standard known as AAC. It is an extension of MPEG-2 audio layer, with refinements for surround sound signals. It has higher coding gains, and many possible channels. It also can use sampling frequencies between 8 kHz and 96 kHz. It has advanced features such as temporal noise shaping, a high level of quantization finesse, and the usage of a plain modified discrete cosine transform. It will be used in the MPEG-4

standard for video objects however it is not yet used in broadcasting applications and is beyond the scope of this book.

4.13.4 Principles Behind MPEG Layer II Coding

The MPEG audio standards use psycho acoustical techniques to reduce the amount of data generated when an analog audio signal is converted to a digital format. The standards take advantage of the fact that a sound element at any particular frequency will have a masking effect on lower level sounds at nearby frequencies. This allows the encoder to reduce the data required to accurately encode an audio signal. This is because any sound elements that are present in the audio signal, but would not be heard even if reproduced exactly, are not coded. The quality of the reproduced signal is very close to that of a compact disc. By exploiting psychoacoustical phenomena such as spectral and temporal masking [8], these coding schemes reduce the irrelevancy of the reproduced audio signal. Another mechanism used to compress the audio signal is the usage of statistical correlations within the audio signal itself.

4.13.4.1 Properties of Pulse Code Modulated Signals

PCM signals are generally used to represent audio and video waveforms digitally. The analog waveform is sampled (measured) at discrete moments in time. The sampling rate or sampling frequency for MPEG-1 audio (as mentioned in Section 4.13.1) can be either 32, 44.1 or 48 kHz. In Section 1.4, earlier, it was shown that one of the properties of a PCM system is that the overall bit rate of the newly generated digital signal is simply the product of the sampling rate and the number of bits allocated to each sample value. This fact also results in a fundamental property of PCM audio, which is that the signal-to-noise ratio of the PCM channel can be determined by selecting a suitable word length for each sample value of the audio signal. The word length of PCM audio is based on the dynamic range required and is generally constant with frequency. Figure 4.5 presents a functional block diagram of an MPEG-1 layer encoder. The input signal can be either a mono or a stereo audio PCM (pulse code modulation) signal.

4.13.4.2 Use of the Fast Fourier Transformation (FFT)

The Fast Fourier Transformation is a widely used digital signal processing technique used in digital broadcasting. Because the FFT can be used in real time it can allow an encoder to calculate the masking effect of dynamically varying audio signals. One of the first functions applied to

Mono or stereo digital audio PCM signal
sampling rate 32, 44.1 or 48 kHz

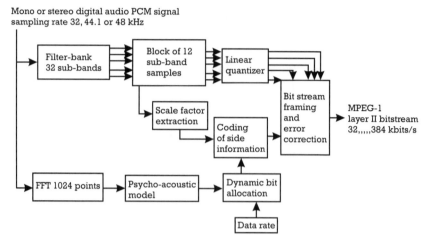

Figure 4.5   Block diagram of an MPEG-1 layer II encoder.

the PCM signal is a Fast Fourier Transformation (FFT) to calculate an estimate of the actual time dependent masked threshold. For the reason
of simulating both spectral and temporal masking effects, a psychoacoustical model is employed after the FFT using fixed rules.

4.13.5 Sub-Band Coding

In parallel to this FFT process, sub-band coding of the PCM signal takes
place. This is a widely used process that takes advantage of the fact that
real sounds do not have uniform spectral energy. When a signal with a
nonuniform spectral density is conveyed with PCM, the whole dynamic
range is occupied only by the loudest frequency component of the signal.
As a result the other signal components are encoded with excess head-
room. Sub-band coding in this case splits the audio signal into 32 frequency bands, and compands each band in proportion to the audio level
found within that band. As a result bands in which there is very little
energy result in small amplitudes that can be transmitted with shorter
word lengths.

The sub-band samples are quantized and coded with the intention to
keep quantization noise below the masked threshold. Layer I uses a block
companding technique with a scale factor consisting of 6 bits, which is
valid for a dynamic range of about 120 dB and a blocklength of 12 sub-
band samples.

The output of the quantizer, the scale factors, and the value of the number of bits per sample are processed into a bit stream. After encoding of the audio signal an assembly block is used to frame the MPEG audio bit stream, which consists of consecutive audio frames. The frame length of layer corresponds to 1,152 PCM audio samples. Each audio frame starts with a header followed by bit allocation information, scale factor and the quantized and coded sub-band samples. The header contains 12 synchronization bits and 20 bits for system information. Optionally, a part of the header and the number of bits per sample can be protected by means of error correction data (16 bits).

4.13.6 MPEG Audio Decoding

The MPEG audio decoding systems for layer I and layer II are very similar and can be simply considered as the reverse of the encoding process. As can be seen in Figure 4.6 below, error correction information is first decoded in order to detect and correct any errors. Next the sub-bands are regenerated. This is achieved by using the values of the decoded scale factors and the decoded value of the allocated number of bits per sample. By using an inverse filter bank the sub-bands can be reconstructed into the digital information signal. By using a standard digital to analog converter (D/A) the audio signal can finally be generated.

4.13.7 DVB Guidelines for Audio Coding

The DVB project has taken a sensible decision in recommending the use of the MPEG system for audio as well as video in broadcasting applications. The main mandatory guidelines from DVB are available in an implementation guidelines document on the topic of digital audio coding. Some of the more important points are summarized as follows:

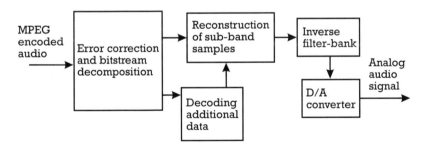

Figure 4.6 Functional description of an MPEG (mono) decoder, for layer I and layer II.

1. MPEG-2 layer I and layer II are supported by the IRD (integrated receiver decoder).

2. IRDs support single-channel, dual-channel, joint stereo, stereo, and the extraction of at least a stereo pair from MPEG-2 compatible multichannel audio.

3. Sampling rates of 32 kHz, 44.1 kHz, and 48 kHz are supported by IRDs.

4. The encoded bit stream does not use emphasis.

5. The use of layer II is recommended for the encoded bit stream.

References

[1] Watkinson, J., *Compression in Video & Audio*, Focal Press, 1995.

[2] Ely, S. R., "MPEG Video, A Simple Introduction," *EBU Technical Review*, No. 266, 1995, pp. 12–23.

[3] ISO/IEC, 13818-2, "Generic Coding of Moving Pictures and Associated Audio Information: Video."

[4] Watkinson, J., *The Engineers Guide to Compression*, Snell & Wilcox, 1996.

[5] ISO/IEC, 11172-3, "Coding of Moving Pictures and Associated Audio for Digital Storage Media at up to 1.5 Mbits/s: Audio Part."

[6] Dehery, Y. F., et al., "MUSICAM source coding for digital sound," *Proc. 17th Int. Television Symposium*, Montreux, June 1991, pp. 612–617.

[7] ISO/IEC, 13818-3, "Coding of moving pictures and associated audio: audio."

[8] Stoll, G., "MPEG Audio Layer: A Generic Coding Standard for Two Channel and Multichannel Sound for Digital Video Broadcasting, Digital Audio Broadcasting and Computer Multimedia," *EBU Digital TV Forum*, Prague, Sept. 25–26, 1995.

MPEG-2 Transport Streams

5.1   Introduction

In the earlier chapters we have seen how an MPEG-2 video encoder could be used to generate different bit rate signals depending on the bit rate required for the application. We have only examined the concept of encoding one single program service, however as we will see in this chapter with digital television (DTV) technology it is possible to combine multiple MPEG-2 encoded program services for the purpose of DTV multicasting, that is, broadcasting multiple DTV services on the same radio frequency broadcast channel.

In this chapter we will see how the MPEG-2 specification can be used to generate so-called MPEG-2 transport streams (TS) containing multiple packetized MPEG-2 elementary streams (PES) of video, audio, and data from different sources. The more commonly used abbreviations will be explained and defined. An emphasis will be placed on building MPEG-2 transport streams using electrical interfaces that are compatible with transmission systems for broadcasting on terrestrial and other distribution networks to the home. It should be remembered that MPEG refers not only to a process of compressing digital audio and video, but also to a means of packaging these signals together with other sources. The standard defines the bitstream that results from encoding and multiplexing and not the actual process of generating the bitstream. This means that it should not need to be revised as new efficiencies are achieved in compression techniques.

Finally constant bit rate and variable bit rate sources will be examined. Important features such as statistical multiplexing and opportunistic data will be discussed in the context of DTV multicasting. The key to DTV multicasting is the use of packets of digital information, containing video, audio, and data. This allows the broadcaster to change the mix of these components dynamically, to increase the efficiency of information transfer to the consumer over a bandwidth-limited channel.

5.2 The MPEG-2 System Layer

The MPEG-2 system layer includes a mechanism for combining MPEG encoded video and audio with associated data signals into transport streams [1]. It can then also allow the combination of any number of different programs into a single MPEG-2 transport stream. This is achieved by first multiplexing the individual components of a television signal together, and then multiplexing many television signals onto one MPEG-2 TS.

5.2.1 Clock Reference Information

A very important part of the system layer is the clock reference information, which is used to enable the recovery of individual video and audio clocks. With MPEG-2 encoding there can be variable delays experienced for video and audio signals due to variable length coding. The system layer includes decoding time stamps (DTS) and presentation time stamps (PTS) which help to ensure the delay is consistent, and that the video, audio, and data are kept in synchronization even though the signals are packetized [2].

5.2.2 Transmission and Program Information

The system layer also provides useful information about the actual transmission network itself (Network Information Table), about the programs being carried (Program Allocation Tables), and finally Conditional Access Tables which reference scrambled or encrypted services. These facilities allow a decoder to select a particular television service from the bouquet of services being delivered to the consumer at any particular time.

5.3 MPEG-2 Bit Streams

As an aid to understanding the different MPEG-2 bit streams a conceptual single service MPEG-2 TS multiplexer is shown in Figure 5.1 below. Here the video, audio, and data elements are encoded and packetized separately. Then the resulting packets are combined or multiplexed together to form a single MPEG-2 transport stream.

Elementary streams (ES): These are the most basic type of MPEG-2 signals and are generated by the encoder. They are separate data streams for video, audio, and data signals.

Packetized elementary streams (PES): These variable length packets of data also contain time stamps and header information. They are separate data streams for video, audio, and data signals. The structure of an MPEG-2 PES is shown in Figure 5.2.

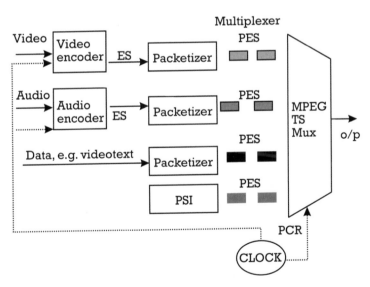

Figure 5.1 An MPEG-2 TS multiplexer.

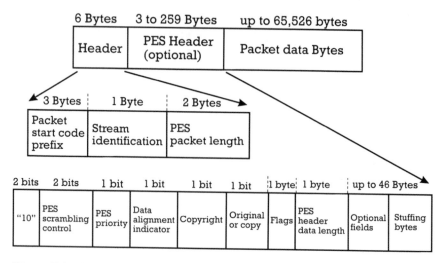

Figure 5.2   MPEG-2 PES structure.

Program stream (PS): This is a multiplex of (audio, video, and data) packetized elementary streams using the same reference clock. It is a simple way of combining the information signals from one television service together. This program stream is only ever intended for recording purposes or for transmission over secure channels that do not introduce errors. The PS has variable length packets. It is not as common a format as the transport stream (see below), and rarely encountered with transmission equipment.

Transport stream (TS): This is the most commonly found format in applications and a typical MPEG-2 TS packet structure is shown in Figure 5.3 (with the optional adaptation field). It can contain one or more television services multiplexed together with independent clocks. Transport streams have fixed length packets (usually 188 bytes) and are intended for transmission over error prone channels (real transmission channels!). This is the format which is used as input to broadcast channel encoders (refer to Chapter 7) prior to transmission to the home receiver. It also contains the transport stream header, which is a group of four bytes of data at the beginning of each MPEG-2 TS packet containing the data required (termed PID) for the demultiplexer to identify programs within that particular TS. By using null packets within the transport stream it is possible to ensure that a fixed rate transport stream is generated from varying bit rate program streams.

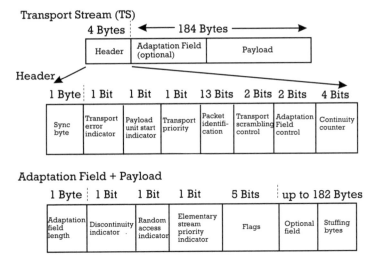

Figure 5.3 MPEG-2 TS packet structure.

5.4 MPEG-2 Interfaces

The two most commonly used interfaces to interconnect MPEG-2 equipment for transmission systems are the:

1. MPEG-2 TS SPI (Synchronous Parallel Interface);

2. MPEG-2 TS ASI (Asynchronous Serial Interface).

These are the two standard interfaces most widely used in equipment and are recommended for all transmission applications. It should be noted that there are many other interfaces used to interconnect electrical equipment carrying MPEG-2 signals. As a result it has often been found that MPEG-2 equipment using older nonstandard interfaces can be difficult to interconnect to newer equipment.

5.4.1 Synchronous Parallel Interface

The Synchronous Parallel Interface (SPI) is commonly used to interconnect MPEG-2 signals to modulation equipment including television transmitters at headends and mainstation sites over very short distances.

The most common variant of this interface is the low voltage differential signal (LVDS). It is a balanced line driver interface which carries 2 sets of 8 data lines, 2 clocks, 2 ground connections, 2 data valid lines, 2 packet synchronization lines, and a cable shield line as shown below in Table 5.1. As this is a balanced line interface, interconnecting any two pieces of equipment can be conceptualized as connecting a "transmitter" unit and a "receiver" unit (see Figure 5.4).

The data packets are sent on two separate channels (A and B), which are balanced. This interface uses a 25 pin D type subminiature connector to connect 11 line drivers and 11 line receivers from a source to a receiver (see Figure 5.5). It is not intended for long cable runs (maximum 10 meters in practice) and is not preferred for harsh environments where rough treatment may be expected, as the cable and pins are easily damaged. The signals and pin outs are shown in Table 5.1 below.

The clock and data timing references for this interface are shown in Figure 5.6.

The maximum output impedance with this system is 100 Ω.

The frequency of the clock signal (Fc) depends on the MPEG-2 TS data rate (Fd). The clock period (T) is given by the expression $T = 1/Fc$, and the clock pulse width (t) is given by $t = T/2 \pm T/10$. The data hold time Td is the same as the clock pulse width.

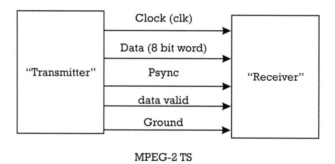

Figure 5.4   MPEG-2 SPI using the LVDS interface.

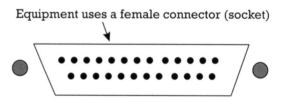

Figure 5.5   MPEG-2 SPI 25 pin type D connector.

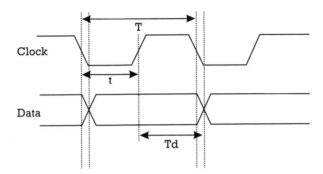

Figure 5.6 MPEG-2 SPI (LVDS) clock and data time references.

$Fc = Fd/8$ (without Reed Solomon error correction)

$Fc = (204/188) \times Fd/8$ (with Reed Solomon error correction).

5.4.2 Asynchronous Serial Interface

The Asynchronous Serial Interface (ASI) is a subset of a well-known ANSI standard (X3T11), and is commonly used in the interconnection of digital transmission equipment. It uses a single coaxial BNC connector and has a 75Ω impedance. It is similar to the SPI connection in that equipment uses a female (socket) connector and cable uses a male (plug).

Table 5.1
MPEG-2 TS, LVDS SPI

Pin number	Signal line	Pin number	Signal line
1	Clock A	14	Clock B
2	System ground	15	System ground
3	Data 7 (A)	16	Data 7 (B)
4	Data 6 (A)	17	Data 6 (B)
5	Data 5 (A)	18	Data 5 (B)
6	Data 4 (A)	19	Data 4 (B)
7	Data 3 (A)	20	Data 3 (B)
8	Data 2 (A)	21	Data 2 (B)
9	Data 1 (A)	22	Data 1 (B)
10	Data 0 (A)	23	Data 0 (B)
11	VAL A	24	VAL B
12	Psync A	25	Psync B
13	Cable shield		

It is gaining popularity among broadcasters, as the physical connection is more rugged and can be used to span longer distances (typically 200 meters). It also has the advantage that the complete signal is contained within a single conductor or cable, as opposed to the 25-wire cable used for SPI interfacing. This makes it easy to handle if automated line switching techniques are required to protect circuits in distribution systems. Many types of transmitter input equipment (TIE) use this standard for routing primary and secondary feeds to broadcast transmitters.

The transmission is bit serial and each 8-bit MPEG-2 TS word is first converted to a 10-bit word. The process is shown in Figure 5.7 below. Then synchronizing fill bytes (K28.5) are added to this new TS, and the resultant TS has an overall transmission rate of 270 Mbps irrespective of the bit rate of the MPEG-2 TS payload. In the ASI receiver these K28.5 bytes are ignored.

5.5 MPEG-2 Transport Stream Parameters

Some of the more commonly used parameters of an MPEG-2 TS are described below. The contents and values of particular tables can be set during the multiplexing of a particular MPEG-2 transport stream containing multiple services.

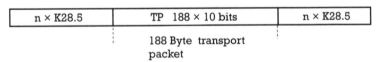

270 MBits/s

K28.5	TP	K28.5	K28.5	K28.5	TP	K28.5

 8 bits   n × K28.5 stuffing   8 bits
               bytes (10 bits)

270 MBits/s

n × K28.5	TP 188 × 10 bits	n × K28.5

               188 Byte transport
               packet

               K28.5 character : 001111 1010
                                 110000 0101

Figure 5.7   MPEG-2 Asynchronous Serial Interface.

AF	The Adaptation Field is uncoded ancillary program data transmitted after the TS header of a data stream.
BAT	The Bouquet Association Table describes a bouquet of programs provided by a broadcaster.
CA	Conditional Access gives information on whether the program is scrambled.
CAT	The Conditional Access Table references a scrambled program.
DTS	Decoding Time Stamp is only transmitted if it has a value different to the Presentation Time Stamp and is used to decode time.
EIT	The Event Information Table is an electronic television guide.
EMM	Entitlement Management Messages are sent with the CAT if a program is scrambled.
ECM	Entitlement Control Messages are sent with the CAT if a program is scrambled.
NIT	Network Information Table gives information about the transmission network broadcasting the digital television service.
PAT	Program Association Table is a list of all the programs contained in the Transport Stream (TS) multiplex with reference to a program map table.
PCR	Program Clock Reference is a reference for the 27 MHz clock regeneration and is transmitted at least every 0.1 second.
PID	Packet Identification is the identification of programs in the transport stream.
PMT	Program Map Table gives program features such as name and copyright, and also references packets to the program clock and references data streams to a program.
PSI	Program Specific Information is data transmitted in the TS for the receiver demultiplexer.

PTS   The Presentation Time Stamp is a time stamp used for vision and sound which is transmitted at least every 0.7 second and it is integrated into the Packetized Elementary Stream.

RST   The Running Status Table is an accurate and fast adaptation to a new program running order if time changes occur in a program schedule.

SI   Service Information is all the information required by a receiver to demultiplex and decode programs in the multiplex.

SCR   System Clock Reference is a reference in the program stream for synchronizing the system-demultiplexing clock in a receiver. It is also integrated into the PES and transmitted at least every 0.7 second.

SDT   Service Description Table gives a description of the programs or services offered in the multiplex.

STC   System Time Clock is a 27 MHz time clock regenerated from the Program Clock Reference for jitter free readout of MPEG data.

SYNC   (Byte) is a synchronisation byte in the TS header.

TS   This is the Transport Stream (see Section 5.3).

TS Header   The first 4 bytes of each TS packet contain the data (PID) required for the demultiplexer in addition to the sync byte. These bytes are not encoded.

TDT   Time and Date Table contain the Universal Coordinated Time (known as UTC) and date.

TOT   The UTC time and date with an indication of the local time offset.

5.5.1 Program Specific Information (PSI)

The MPEG-2 systems standard has specified program specific information (PSI) data to allow the automatic configuration of an integrated receiver decoder (IRD) or set top box (STB). This information is needed to demultiplex and decode the video and audio information carried in the MPEG-2 TS.

The MPEG-2 PSI data is structured as four distinct tables.

1. The program association table (PAT).

2. The program map table (PMT).

3. The network information Table (NIT).

4. The conditional access table (CAT).

5.5.2 DVB Service Information (SI)

Within the DVB specifications there is found another level of service information (SI), which adds to the MPEG PSI by providing data to aid in the automatic tuning of IRDs, and other data intended for display to the consumer [3]. This SI data provides information to the consumer that aids in selection of services and program events. The data contained in the SI can be used as a basis for an electronic program guide (EPG).

DVB has specified seven additional SI tables, of which three are mandatory and four are optional [4]. These tables may be quite large and take many individual transport stream packets to complete. A brief description of these tables now follows:

1. The network information table (NIT) gives information to the set top box or IRD about the network carrying a particular service. This might include the channel frequency to aid the IRD in searching for a particular service. Each network is given a unique identification number. If there are regional opt-outs or variations in service then each will have a unique identity and be considered as a different network.

2. The service descriptor table (SDT) allows the attachment of text names to the services in a multiplex.

3. The event information table (EIT) gives information about the program being carried. A program can be considered as a series of distinct events, with advertisements carried between events. There are two types of EIT. The present/following event (EIT pf) contains information on the present and following program on each service. This tells the consumer what programs are currently available for selection and what program are going to follow the present programs on all services. The schedule (EIT s) data in the EIT provides information for up to 10 days in advance. This feature

allows the consumer to plan the viewing of specific events or programs in advance and have on-screen reminders when the programs are about to start.

4. The bouquet association table (BAT) allows a broadcaster to link services that might be delivered by different methods across different multiplexes. A bouquet is best described as a collection of services marketed as a single entity. When linked to an EPG it can be a powerful way of controlling access to broadcasters' services, across any transmission medium.

5. The running status table (RST) provides data concerning the (running or not running) status of an event. A lot of event information changes at once when there is a program junction. This information could be used to control, for example, the recording of an event using a video cassette recorder.

6. The time and date table (TDT) contains the universal coordinated time (UTC) and date, which is used in timing broadcasts (see Chapters 10 and 14).

7. The time offset table (TOT) provides information concerning the present time and date with a local offset from the UTC.

8. The stuffing table (ST) is used to invalidate other sections of the Service Information. This occurs, for example, when a MPEG-2 TS containing services is remultiplexed to add or remove services.

5.5.2.1 Actual and Other SI

Service information can also be categorized as actual and other. SI actual data refers to the SI data carried by a multiplex and concerning that multiplex itself. SI other refers to the other multiplexes that the receiver is capable of decoding from the same transmission medium. In some countries it is a requirement of digital terrestrial television that each multiplex carry data about all other multiplexes on that transmission medium. SI other causes problems for single frequency networks (SFN) where the SFN service shares the same transmit antenna as other regional services (see Chapter 10). Table 5.2 below gives an overview of SI tables.

Table 5.2
Service Information

	MPEG-2 PSI	DVB SI (Mandatory)	DVB SI (Optional)
Network information	PAT	NIT	NIT
Bouquet information	CAT	—	BAT
Service description	PMT	SDT	SDT
Event information	—	EIT	EIT
Running status	—	TDT	RST
Stuffing	—	—	ST
Time offset	—	—	TOT

5.5.3 Service Information Distribution

The service information being broadcast from different transmission sites in a network will generally be different. This is because different sites will use different broadcast channels for the same service, and because regional differences in services can be expected due to possible program opt-outs at different times of the day. It is also possible to have even a completely different service being broadcast at these times using possibly a different language. If television services are required to broadcast SI other data then it can become very difficult to manage the SI data without a separate SI distribution system and the local multiplexing of SI data onto the MPEG-2 TS prior to broadcasting. This SI other requirement can eliminate the possibility of using a single frequency network (SFN) as the data being broadcast will change from site to site. A requirement for SFN operation is the synchronous broadcasting of the exact same MPEG-2 TS from all sites in the SFN (see Chapter 10).

A typical solution to SI distribution is to have a central collation point for all SI information and then distribute this data to all the final transmitting stations. At these service information insertion points (SIIP) the relevant SI data for a particular multiplex is added to the MPEG-2 TS. Then the new MPEG-2 TS multiplex can be broadcast and passed onto other transmitter sites in the same network. Only one SIIP per multiplex is required, and the SI data is stored in cache memory at the SIIP so that only an update to SI data is required from the central collation point. As SI data can take a considerable amount of time to be downloaded at the

SIIP, using a cache memory system that requires only SI data updates considerably improves the system response time.

5.6 Constant and Variable Bit Rate Sources

As the names above suggest MPEG-2 multiplexing can take account of constant and variable bit rate sources and manage the amount of data that is produced by these sources to match the bandwidth of the transmission channel. This transmission channel can itself have a fixed bandwidth (in the case of a broadcast channel) or a variable bandwidth (in the case of an ATM network).

Most MPEG-2 TS multiplexers have constant bit rate (CBR) sources as inputs and they then multiplex the many programs onto the final TS in prescribed proportions which are closely controlled and fixed. However because compressed video is bursty in nature there will be many times when an MPEG-2 encoder will not need all of the prescribed bit rate due to still or easily predicted picture content. At these times the video bit rate will drop substantially and the multiplexer will add null packets to the TS to make up the bit rate to that previously prescribed. Clearly this is wasteful of system capacity and there are now innovative multiplexing techniques available to use this instantaneous excess capacity. In the following sections some applications of these techniques will be presented.

5.6.1 DTV Multicasting

An MPEG-2 TS multiplexer can be used to combine not only the components of a single television service (video, audio, and data) but also different television programmes or services. This allows a single MPEG-2 TS to carry many different television services to the consumer. This multiple service multiplexing technique is shown in Figure 5.8.

As mentioned earlier in Section 5.1, the key to DTV multicasting is the usage of MPEG-2 packets. These packets of data can be error corrected and switched very quickly using hardware. In multicasting, each program may have its own independent program clock reference (PCR). These independent 27 MHz clocks are not locked together, and this allows the multiplexing (or mixing) of independent programs, from different locations, which may have even been encoded at different times. If an MPEG-2 decoder switches to another program within the same

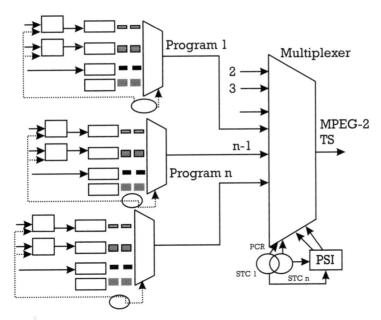

Figure 5.8   Multiple service multiplexing.

MPEG-2 TS bouquet the new PCR samples are quickly acquired from the TS and an internal phase locked loop (PLL) circuit can regenerate the 27 MHz clock for that program.

Packet multiplexing must obey some rules in order to ensure that a program stream has audio and video in proper synchronization. Otherwise it might be possible to send the video packets of an entire program before the audio packets! To avoid this scenario, packets of video, audio, and data are multiplexed onto the final TS in proportion to the instantaneous bit rate of each. This ensures that synchronization between all sources is maintained. This is achieved by scheduling packet delivery using the packet scheduler or transport multiplexer. The packet scheduler monitors the buffer levels of a hypothetical decoder called the transport stream system target decoder (TS-STD). It ensures that none of the many video, audio, and system buffers overflows or underflows.

5.6.2 Statistical Multiplexing

This is a technique for combining multiple digital television (DTV) services onto a single MPEG-2 TS using the dynamic variances in picture content to allocate bit rate among the services in the TS. It is a very exciting development in multiplexing technology that is popular with broadcasters as it allows for the efficient usage of the limited resource (multiplex bandwidth) by multiple services.

Statistical multiplexing or "stat muxing" is a dynamic process that exploits the fact that the different video streams from different programs are usually uncorrelated. When several video streams are simultaneously encoded and multiplexed together to form an MPEG-2 TS, it is generally found that they require different bit rates which depend on the source material content. Thus, for example, a high action picture with a lot of movement and camera changes will generate a much higher data rate stream, than a studio based talk show where there is very little movement within the picture, and from picture to picture. As was shown in Chapter 4, very different compression gains may be achieved for both program types. In Figure 5.9 below an illustration of the required bit rate to encode a video signal is shown to illustrate the dynamic nature of the bit rate requirement. Note that a fixed bit rate encoder will not match the characteristic shape of this curve, but in general it will be a fixed bit rate which will exceed the service-encoding requirement at all times. However with a statistical multiplexing scheme it is possible to match the characteristic requirements of the service to be encoded in real time, hence liberating bit rate at times for other services.

The overall bit rate for an MPEG-2 TS will be fixed by the broadcaster to match the data payload of the channel encoding system used. For example, in terrestrial broadcasting (using the DVB-T system) the mode of operation is chosen and then fixed, within the modulator, by the broadcaster. This sets a maximum usable MPEG-2 TS payload of between 5 and 31 Mbps. This value once set cannot be exceeded, however the challenge then is to maximize the efficiency of usage of this MPEG-2 TS. As the sum total of the services cannot exceed the channel capacity of the transmission system, significant gains can be made over constant bit rate (CBR) multiplexers by allocating more bits to some of the video encoders as they need them, and equally reducing the amount allocated to other video encoders to ensure the maximum sum total is not exceeded. This is a dynamic process and it works by assuming that at any time the program content of different services will be uncorrelated.

Two possible conditions exist which are of interest.

1. If all the services require very little bit rate individually in order to adequately encode them, then it is relatively easy to manage the

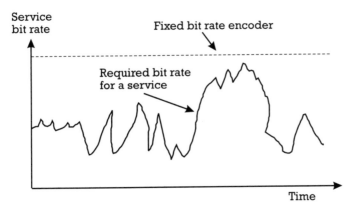

Figure 5.9 Dynamically varying bit rate required by a service.

overall requirement for a constant sum bit rate by allocating null packets to each service to ensure the sum total is constant.

2. If all the services require a large bit rate to adequately encode them at the same instant, then the overall maximum sum total of the TS will be exceeded, and the statistical multiplexer will need to reduce the bit rate allocation to some of the services if not all. This could be achieved by less accurate encoding of the source material. However, if a signal has been MPEG-2 encoded prior to multiplexing with the other MPEG-2 signals it may not be possible to reduce the bit rate of this signal without re-encoding the signal again, which can lead to concatenation error being introduced to the signal.

There are some important issues that are of concern when using statistical multiplexing techniques. If a multiplex has been allocated between a number of service providers, e.g., $\frac{1}{2}$ of a multiplex to service A and $\frac{1}{2}$ to service B, then it may be difficult to obtain agreement to use statistical multiplexing on this MPEG-2 TS, as neither service provider will want to reduce the bit rate allocation to their respective service. Statistical multiplexing may make it more difficult to control regional opt-outs and allow local service insertion. Statistical multiplexing works best in conjunction with a well-scheduled mix of services and more importantly a well-scheduled mix of service content [5]. This avoids the problem of the simultaneous encoding and multiplexing of material from different services, each requiring a high bit rate onto a TS.

5.6.2.1 Stat Mux Gains

Some manufacturers claim that using statistical multiplexing can reduce the required bit rate of a transport stream by between 20–30% at any time. This is achieved by setting a minimum service quality level, not a minimum service bit rate. The output of each encoder is monitored frame by frame and the bit rate allocated to a service is optimized to ensure constant quality. High action scenes are given more bandwidth when required, while static scenes are given less bandwidth. Hence the system can match the required bit rate of the service content. Because most of the time the sum of the required bit rates for all of the services will not exceed the overall transport stream rate, this excess capacity is available for other usages including opportunistic data.

5.6.3 Opportunistic Data

This is another application that uses up the instantaneous excess capacity of an MPEG-2 TS that is generated from the presence of easily compressed images. In this case the excess capacity is given over to download data that may or may not be associated with the program content. This data takes advantage of the instantaneous excess capacity and fills up the transport stream, and as a result is termed opportunistic data. An example of the usage of this opportunistic data is the downloading of an interactive electronic brochure during an advertisement. The electronic brochure in this example is related to the advertisement. With cheaper and cheaper electronic storage becoming available to consumers, it will be possible to slowly download any data material to a home storage device for fast access to the information at a convenient time. Mass storage in set top boxes at low cost will open up opportunities for the provision of new services in limited capacity digital transmission systems. This will be achieved by exploiting the time varying requirements for capacity of digital signals.

References

[1]   ISO/IEC 13818-1, *Generic coding of moving pictures and associated audio information: systems*, 1994.

[2]   Steinberg, V., *Video Standards, Signals, Formats and Interfaces*, Snell & Wilcox, 1997.

[3]   EBU/CENELEC/ETSI-JTC, *Digital Video Broadcasting: Specification for Service Information (SI) in DVB systems*, ETS 300 468, 2nd ed., January, 1997.

[4]   De Bruin, R., and J. Smits, *Digital Video Broadcasting, Technology, Standards, and Regulations*, Norwood, MA: Artech House, 1999.

[5]   Isnardi, M., "DTV Multicasting," *Broadcast Engineering*, September, 1998.

CHAPTER 6

Contents

Single Carrier Digital Modulations

6.1 Introduction

In this chapter digital transmission technologies will be discussed. Emphasis will be placed on digital channel coding (referred to as modulation) methods that are used in single carrier broadcasting systems. In a later chapter multicarrier modulation techniques such as coded orthogonal frequency division multiplexing (COFDM) will be treated in detail. Single carrier systems use, as the name suggests, only one carrier signal onto which the information signal is modulated. These modulation methods dramatically decrease the bandwidth required to transmit a digital signal in a radio frequency (RF) channel. As was seen in earlier chapters, compression may be used to reduce the bit rate of a digital signal representing video or audio information. This chapter does not have anything further to add to the digital compression techniques already described, but seeks instead, to describe the more commonly used digital transmission schemes that map data onto RF channels in as bandwidth efficient a manner as possible. It will describe single carrier modulations as are now used in satellite, cable, MMDS (microwave multipoint distribution systems), and some terrestrial transmission systems. However the DVB terrestrial modulation standard (known as DVB-T) is based upon the use of a multi-carrier modulation scheme (COFDM), and as mentioned above this will be dealt with in considerable detail in Chapter 7. However it should be noted that the inner modulations used with the DVB-T standard are common to single carrier modulations and therefore the reader is advised to study this chapter before reading Chapter 7. The more commonly used modulations for digital transmission include quaternary phase shift keying, hereafter referred to as QPSK (or 4-QAM), 16 level quadrature amplitude modulation (16-QAM), 32-level quadrature amplitude modulation (32-QAM), 64-level quadrature amplitude modulation (64-QAM), and vestigial sideband modulation (VSB). These modulations will be described in the subsequent sections of this chapter.

6.1.1   Uses of Single Carrier Digital Modulations

Single carrier digital modulations are used widely in different transmission media such as MMDS, cable, terrestrial, and satellite networks. QPSK modulation is used in satellite transmission systems such as the direct to home (DTH) digital broadcasting system using the DVB-S system [1]. This is because the modulation is rugged and performs well with a low carrier to noise ratio (C/N), and also because QPSK modulation gives the best performance of all available modulation types in a power limited transmission system such as a solar powered satellite transmitter.

64-QAM modulations are used in cable and some MMDS networks [2] because the data capacity of the modulation is high and because in this case transmitter power is not limited. The required C/N ratio for error free reception is higher, but the carrier level can be readily boosted by amplification throughout the transmission medium.

8-VSB systems are used in some terrestrial networks because of their high data capacity in a multipath transmission medium [3].

6.2   Bandlimiting Digital Signals

For the purposes of transmission, a baseband digital information signal can be visualized as a sequence of pulses, which is similar in appearance to a square wave signal. It can be shown mathematically that any non-sinusoidal, repetitive, single-valued waveform consists of sine waves

and/or cosine waves. The frequency of the lowest frequency, or fundamental, sine wave is equal to the repetition rate of the nonsinusoidal
waveform. All of the other signals are harmonics (multiples) of the fundamental and there are an infinite number of these harmonics. A typical
digital information signal can be represented by such a waveform.

For example, a square wave with a repetition rate $\omega/2\pi$ per second,
and amplitude A, may be represented using Fourier analysis by a time
varying voltage $E(t)$, where

$$E(t) = \frac{4A}{\pi}(\cos \omega t - 1/3 \cos \omega t + 1/5 \cos 5\omega t - 1/7 \cos 7\omega t + \ldots) \quad (6.1)$$

Notice that the fundamental component, $\cos\omega t$, has the largest amplitude and that the sequence is infinite. A digital information signal will
have a similar characteristic to the above example.

6.2.1 Digital Signals in Time and Frequency Domains

A digital information signal consists of a series of binary logic values,
zeros, and ones. These ones and zeros are represented electrically by a
sequence of digital pulses, each of which has a finite duration of T seconds. By using Fourier analysis it can be shown that any idealized digital
pulse that has a finite duration of T seconds will have an infinite spectrum in the frequency domain! The converse unfortunately is also true,
any pulse with a finite frequency spectrum will have an infinite duration
in the time domain. This implies that a pulse with a finite frequency spectrum will have a corresponding infinite time duration waveform, and
therefore consecutive pulses will overlap with each other and cause inter
symbol interference (ISI). This means that these pulses would not be suitable for use in a real transmission system, because every pulse would
interfere with every other pulse. As a result of this fact bandlimiting is
required for pulses in the time domain. This can be achieved with the
usage of filters. Note that symbols are equated to pulses at this time, for
reasons that will become apparent later in this chapter.

6.2.2 Nyquist Filtering

Nyquist filters are theoretical filters, named after the twentieth-century
electrical engineer Nyquist. They are used to describe the process needed
to limit the bandwidth of a pulse sequence prior to modulation. They are
useful starting tools in modelling the characteristics of the filters needed
to limit the pulse spectrum. A Nyquist filter has an idealised rectangular
frequency response and zero group delay. According to the Nyquist

(pulse shaping) criterion, the bandwidth (BW) occupied by a pulse spectrum after filtering is;

$$BW = (R_s / 2)(1 + \alpha) \qquad (6.2)$$

In (6.2) R_s represents the symbol rate (or pulse rate), and α is the filters roll-off factor, where $0 < \alpha < 1$ [4]. When $\alpha = 0$ the filter is known as a Nyquist filter and it has a rectangular frequency response and zero group delay. The filter bandwidth is exactly half the symbol rate. Here pulses are termed as symbols, however symbols will be seen later in this and other chapters to represent more than just one pulse. In Figure 6.1 a finite pulse is shown (a) together with a Nyquist filter (b) and the resultant output waveform (c). It can be noted that the resultant signal waveform after the filtering process has a duration much in excess of that of the original pulse signal, and that the waveform has zero amplitude at intervals of T, the pulse duration. All subsequent pulses will have the same characteristic waveform in the time domain, with zero amplitude at intervals of T. If the receiving equipment samples the pulses at these points T, $2T$, $3T$, and so forth, it is possible to decide on whether a one or a zero was transmitted. It should be noted that the Nyquist filter is an idealized filter and is therefore not practicable to manufacture.

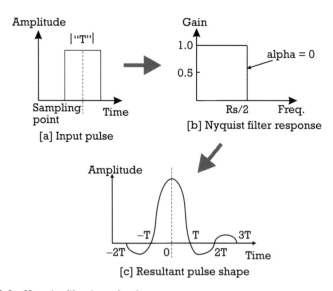

Figure 6.1   Nyquist filtering of pulses.

6.2.3 Raised Cosine Filters

If α is set to 1 in (6.2) above, the filter response formed is that of a so-called raised cosine filter. The filter has a frequency response that is skew-symmetric about a frequency point equal to half the symbol rate. The resulting pulse shape after the filter is similar to that of the Nyquist filter, except that the resulting pulse has less amplitude outside the main pulse period. Hence the response is better than that achieved with a Nyquist filter but the filter has a larger bandwidth. In Figure 6.2 the typical raised cosine spectra are shown with that of a Nyquist filter ($\alpha = 0$).

Real filters used prior to modulation, are usually raised cosine filters with roll-off factors or values typically in the range 0.1 to 0.4. The bandwidth of a signal derived from such filters will be equal to the symbol rate multiplied by $(1 + \alpha)$.

6.3 Data Transmission Speeds

The rate of data transfer through a communications channel depends on many parameters including the signalling speed. The signalling speed of a communications channel is often referred to as the baud rate. A baud is a measure of the signalling speed and in a system where all pulses (or symbols) have the same duration, the speed in bauds is equal to the maximum pulse transmission rate. This is not the same as the information bit

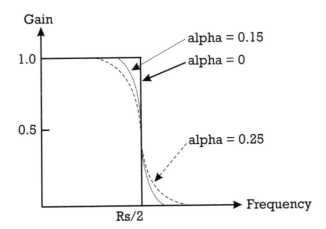

Figure 6.2   Raised cosine filter response.

rate. If the system used allocates one information bit only to each pulse then the baud rate and information bit rate will be equal. However in systems which encode the information data in such a way that more than one information bit can be placed onto a signalling pulse, the information bit rate will be greater than the symbol or baud rate. Nyquist determined that for an ideal communications channel, the maximum signalling speed (in bauds) of the communications channel is equal to twice the bandwidth of that channel. Hence there is a limitation to the symbol rate that can be transferred in a real channel. However as we will see by using multilevel encoding it is possible to associate more than one information bit to each symbol and therefore increase the amount of data transfer for a fixed maximum symbol rate. The Hartley law gives a useful equation that relates the total channel capacity to the channel bandwidth and number of encoding levels, in the total absence of noise [5],

$$C = 2(BW).Log_2 n \tag{6.3}$$

where C = the channel information capacity (bps), BW = channel bandwidth (Hz), and n = number of encoding levels.

6.4   Multilevel Digital Modulations

There are many ways of modulating digital information onto a carrier signal, but these can be divided into three generic types which include, amplitude shift keying (ASK), frequency shift keying (FSK), and phase shift keying (PSK). The terminology of "shift keying" relates back to telegraphy systems where a mark or a space code is transmitted by "keying" or changing a transmission parameter. For example in frequency shift keying, the nominal unmodulated carrier frequency corresponds to a mark condition, and a space is represented by a downward frequency shift in the frequency of the carrier signal. With phase shift keying the phase angle of the carrier signal is phase shifted by +90° for a mark and by −90° for a space. With four-phase, or quartenary phase shift keying, there are four possible phase shifts, +135°, −135°, +45°, and −45°, which allows the possibility of assigning 2 bits of information to each phase state (or symbol). Most current systems use a mixture of ASK and PSK, where data is transmitted on both the in-phase (I) and the quadrature (Q) components of the carrier signal. These signals are termed as multilevel modulations because there are a number of permitted amplitude levels, and four phases permitted for the carrier signal [6].

The more commonly used multilevel modulations include quadrature amplitude modulation (QAM) and vestigial sideband modulation (VSB). Quartenary phase shift keying (QPSK) allows only two equal amplitude levels and is not strictly a multilevel modulation. Note that some reference sources term QPSK as quadrature phase shift keying, there is no difference in the actual modulation type; however for consistency this book will term QPSK as quartenary phase shift keying. Similarly quadrature amplitude modulation (QAM) is occasionally termed as quartenary amplitude modulation. All of these modulations are used in digital transmission systems over different broadcast media, including satellite direct to home (DTH) broadcasts, cable networks, MMDS systems, and terrestrial systems. Each modulation type has slightly different properties that make that system more attractive for usage on a particular broadcasting medium. There is always a trade-off between the error performance required and minimum data payload needed, which makes some modulations more attractive to particular broadcast media.

6.4.1 QPSK (Quartenary Phase Shift Keying)

As mentioned earlier in this chapter, with this particular modulation the phase of a carrier can be changed to one of four possible phase states depending on the value of the information data bits. Each of the four phase states are 90° apart and since there are four choices for the carrier phase this means that each symbol can represent 2 bits of information (see Equation 6.4). The data information can therefore be 00, 01, 10, or 11.

In the QPSK modulator an input serial data stream is first converted from serial to a parallel bitstream, as shown in Figure 6.3. This process effectively demultiplexes the original serial data stream into two data streams (termed I and Q) with alternate bits from the original data signal on each stream. These two parallel data streams are next low pass filtered to band limit each symbol (as described in Section 6.2 above), and then each stream is double sideband suppressed carrier modulated with a local oscillator (LO) to modulate the data symbols onto the two carrier signals. The local oscillator that is used as the carrier signal is the same signal for both data streams, except that it has a 90° phase shift applied to one of the streams relative to the other stream, hence the use of the terminology I (In-phase) and Q (quadrature phase). This 90° phase shift in the carrier phase ensures that the two resulting signals are orthogonal to each other

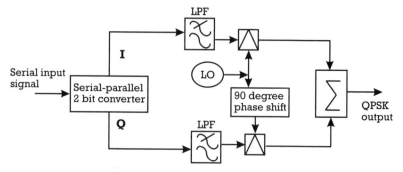

Figure 6.3   QPSK modulator.

and this fact allows the two signals to share the same bandwidth yet be easily separated in a receiver. Differential QPSK, referred to as DQPSK, is used in DAB systems and will be discussed in Section 13.9 later.

The main constraint on a QPSK receiver is that to successfully decode the data information contained in the modulated signal it must be able to determine the absolute phase of the carrier (termed coherent demodulation). As a result of this constraint a learning pattern must be transmitted to the receiver to determine the actual phase of the carrier, or else differential coding must be used.

When viewing the time domain representation of a QPSK (or QAM) signal it is useful to use a constellation diagram. These can be easily generated using an oscilloscope, plotting the I and Q components of the signal against each other (using the oscilloscope *XY* function), refer to Section 9.3. With the constellation diagram of a QPSK/QAM signal each data symbol is represented by a point that shows the peak amplitude and phase of the transmitted data symbol in real time. There are four possible phase/amplitude states associated with a QPSK symbol (see Figure 6.4) and the number of bits that can be assigned to each symbol in this case is 2.

6.4.2   Data Capacity of 2^m-QAM Systems

As mentioned above QPSK and QAM systems can be represented graphically quite easily using constellation diagrams. These diagrams provide a useful means for personnel to visually recognize and distinguish between the different so-called 2^m – QAM systems. Depending on the type of modulation scheme chosen for transmission purposes by the broadcaster, the different modulation schemes can offer significantly different payload

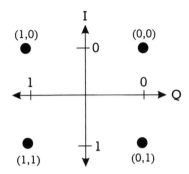

Figure 6.4   QPSK constellation diagram.

capacities. The general rule is given in (6.4) that relates the modulation type to the number of data bits each symbol (or constellation point) can carry.

$$2^m = q \qquad (6.4)$$

Here q is the QAM order, e.g., 64 for 64-QAM, 4 for QPSK, and so forth, and m is the number of data bits that each symbol can carry.

In Table 6.1 below, the data capacity of a symbol is given for different QAM modulation types. As expected the higher order modulations such as 64-QAM have the greatest data capacity per symbol. However as we will see the higher order modulations are less rugged and more sensitive to noise impairment. The highest order QAM system thus far developed is the 256-QAM system, which is not widely used for broadcasting signals to the home, but is used in conjunction with trellis coding in point to point microwave digital link systems. It has a data capacity of 8 bits/symbol but is most sensitive to the quality of the link path and adversely effected by multipath propagation. As the broadcaster has little control over the path between the transmitter and the consumer's equipment, 256-QAM is not regarded as a suitable broadcast modulation.

6.4.3   Byte to M-Tuple Conversion

Depending on the mapping efficiency of the particular modulation chosen, an integer number of 8-bit bytes, will be mapped onto n symbols during the modulation process. The relationship between, n the number of symbols that are formed from k data bytes for a particular modulation type (denoted by 2^m – QAM) is given by (6.5) and shown in Table 6.2.

Table 6.1

Number of Data Bits/Symbol for QAM Systems

Number of data bits per symbol (m)	Modulation type
2	QPSK (4-QAM)
4	16-QAM
5	32-QAM
6	64-QAM

Table 6.2

Relationship Between 2-QAM Modulation and Mapping of Symbols

2^m-QAM	m bits	n symbols	k bytes
QPSK	2	4	1
16-QAM	4	4	2
64-QAM	6	4	3

This process is known in the DVB system as byte to m-tuple conversion [7].

$$8k = n.m \qquad (6.5)$$

6.4.4 16-QAM

With this form of quadrature amplitude modulation there are more possible carrier amplitude levels available for each symbol to be mapped onto. By modulating both the phase and the amplitude of the carrier signal it is possible to create many more constellation points than with the simple QPSK modulation scheme. The 16-QAM signal has, as the name suggests, 16 different possible phase and amplitude combinations. The signal is produced in a similar manner to the QPSK signal using a quadrature modulator. During the modulation process each of the two demultiplexed data streams can be mapped onto one of four different amplitude values depending on the bit order of the digital data signal. With 16-QAM modulation the same double sideband suppressed carrier modulation process is used for the I and Q signals, as is used in a QPSK modulator. A description of the implementation of higher order QAM systems is given in Section 6.4.7.

The 16-QAM constellation diagram has 16 different points, each one corresponding to a particular symbol, and the number of bits associated with each symbol is four, as shown in Table 6.1. In Figure 6.5 below, a 16-QAM signal constellation is shown along with the corresponding data bits of each symbol. The two most significant bits (MSB) of each symbol are denoted as $I_k Q_k$ and express the I and Q component values. These two MSBs represent the quadrant within which the symbol is situated. With the DVB cable specification [7] the 2 MSBs are differentially encoded and afforded high error protection as they are the most important bits of the overall symbol. The least significant bits (LSB) of the symbol are used to differentiate the symbols within a quadrant and are less important. The constellation points in quadrant 1 can be changed to quadrants 2, 3, and 4 by rotation of the LSBs according to Table 6.3 and by changing the values of the MSBs, that is the value of the indices $I_k Q_k$. The same differential coding process is used on the two MSBs of symbols in DVB higher order QAM modulation systems, such as 32-QAM, and 64-QAM.

6.4.5 32-QAM Modulation

This is a particularly interesting modulation format because of the unique constellation diagram that is formed for this variant of a 2^m – QAM system. In the last section dealing with 16-QAM, there were two steps of amplitude and two phase states possible for each of the carriers (I and Q), which resulted in 16 possible symbols ($2^4 = 16$).

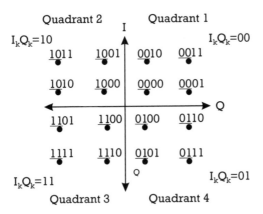

Figure 6.5 16-QAM constellation diagram.

Table 6.3
MSB's and LSB's Relation to Rotation Angle, in QAM Constellation Diagram

Quadrant number	Most significant bit, $I_k Q_k$	Least significant bits rotation
1	00	0
2	10	+90°
3	11	+180°
4	01	−90°

By extending this to three possible amplitude states, it allows the generation of six possible states for each carrier and hence in total there are 36 possible combinations. However for 32-QAM there is only a requirement for 32 combinations or symbols, as each possible 5-bit word combination can be provided for by using only 32 symbols, $(2^5 = 32)$.

6.4.5.1   32-QAM Constellation

A constellation diagram for a 32-QAM system is shown in Figure 6.6. It is worth noting that there are four missing corner symbols that could have been used if a 36-QAM system was desired. However as inferred from the last paragraph, 36-QAM is not used because it is not possible to map an integer length binary word into 36. As a result the 4 most extreme symbols on the corners of the constellation plane are disregarded and not used. These corner states in all QAM systems represent the highest amplitude points for the possible symbols, and would require the transmitter to generate the maximum power level to represent these symbols. By eliminating these symbols the transmitter can reduce power while retaining the ability to broadcast 5 bits of information with each transmitted symbol [8]. 32-QAM is not commonly deployed in broadcast networks at present.

6.4.6   64-QAM Modulation

This is the most widely used modulation format on cable systems and also within the DVB-T COFDM system (see Chapter 7) because of the high data capacity it can afford. It is similar to the earlier described modulations and uses quadrature amplitude modulation techniques to map data bits onto two orthogonal phase carriers (I & Q, as before). The number of data bits carried per symbol is 6 with this particular modulation, the

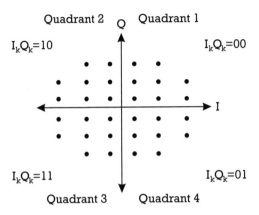

Figure 6.6   32-QAM constellation diagram.

highest of the DVB QAM systems, as shown in Table 6.1. The previously described mapping process is illustrated in Table 6.2, and the rotation of LSBs from quadrant to quadrant is detailed in Table 6.3. A typical 64-QAM constellation is shown below in Figure 6.7, illustrating as in earlier figures, that the two MSBs are differentially coded and used to signal the quadrant position of the symbol data. The subsequent 4 data bits (LSBs) are mapped in accordance with the rotation rules detailed in Table 6.3 above.

6.4.7 QAM Modulator Implementation
Hardware implementation of a QAM modulator is shown in Figure 6.8. It is very similar to the implementation shown for a QPSK modulator,

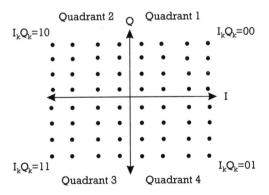

Figure 6.7   64-QAM constellation.

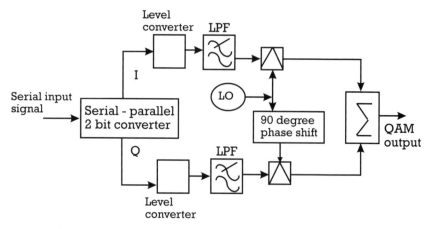

Figure 6.8   QAM modulator.

except that in this case a level converter is used to take groups of bits and translate them to specific voltage levels. This is because higher order QAM systems allow for amplitude variations within each quadrant and depending on the specific pattern of incoming bits to the I and Q branches of the circuit of the modulator, they are translated onto particular voltage amplitudes.

6.4.8   Filtering of QAM Systems
Before modulation the I and Q signals are normally filtered using a square root cosine filter (similar to that described in Section 6.2.3) with a roll-off factor (α) of typically 0.15 (for DVB cable systems). Also the filter should have a low in-band ripple of no more than 0.05 dB, and a linear phase response. For a more detailed discussion the reader is recommended to refer to the DVB specifications and other related documentation [4].

6.4.9   Trellis Code Modulation
Trellis code modulation (TCM) can be thought of as a special form of QAM that adds an extra bit to the n-bit symbol sequence to create twice as many signal constellation points as are required to represent the data. Doubling the number of signal points in this manner ensures that consecutive symbols fall far enough apart to be easily distinguished. In the demodulator the trellis decoder compares the received symbol with all possible valid symbols, and selects the most likely symbol. As a result

errors may be reduced, and TCM modems can operate at higher speeds than non-TCM modems.

6.5   DVB-based QAM Systems

The above-mentioned QAM systems are generic in nature, and depending on actual system implementations they will have incorporated to a greater or lesser extent the features described earlier. Most standards in use today differ slightly in the usage of inner coding techniques and the choice of filter roll-off factor. However, the main features of single carrier QAM/QPSK modulators are as described above. As an example of the main implementation features of a commonly used modulator, a brief description of the DVB-C concepts will be given, however the reader is recommended to refer to the standard for a more detailed description [7].

6.5.1   DVB-C System Concepts

The digital video broadcasting project has defined a standard for the transmission of digital television on cable networks, known as the DVB-C system. The DVB-C specification has been ratified by ETSI, and is beyond the scope of this book. However, a block diagram of a DVB-C modulator is shown in Figure 6.9 below illustrating some of the more important functions. The first function is to adapt the data structure of the incoming MPEG-2 TS (transport stream) to that required by the modulator this process is described in more detail in Section 7.8 and illustrated in Figure 7.4. This may involve forming a 188-byte structure for the MPEG-2 TS from a 204 byte structure. Next the sync 1 byte is inverted in accordance with the MPEG-2 framing structure and the data signal is randomized by passing the signal through a pseudo random binary sequence (PRBS) generator. The PRBS generator ensures that the data is randomly distributed (interleaved) throughout the RF channel spectrum, and this protects against signal fading and ensures that the modulation employed has adequate binary transitions for clock recovery in a receiver. Most error protection codes are designed to combat errors resulting from signal fades if the errors are randomly distributed. The PRBS interleaving process is used to ensure that any error bursts will be randomly distributed.

A concatenated error-protection scheme based on the Reed Solomon error correction code, (and referred to as RS(204,188) or sometimes RS(204,188, $t = 10$)), is used as outer coding to protect the randomized data against up to 8 possible errors per packet. However, in order to

MPEG-2 TS input signal

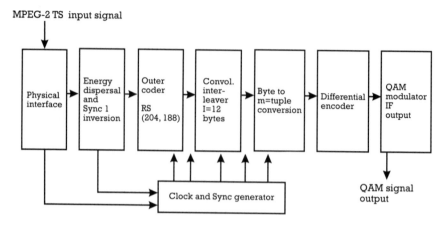

Figure 6.9   Conceptual QAM encoding system.

achieve this level of protection it adds 16 parity bytes to the MPEG-2 transport packet. In the terminology, 188 data bytes have 16 parity bytes added to give 204 bytes in total, and this can then protect against 8 erroneously received bytes per packet.

Next an inner coding scheme is used to protect against a burst or group of more than 8 errors arriving together. To randomize and break up this group of errors, convolutional interleaving is used with (I = 12), conceptually 12 branches. These branches are connected in turn to the input data stream to demultiplex the data onto each branch while maintaining the periodicity of the sync bytes and inverted sync bytes, for synchronization purposes. With a convolutional encoder every encoded bit at the output of the convolutional encoder is a linear combination of some previous bits. The next steps in the process involve byte to symbol mapping, using byte to m-tuple conversion, differential encoding of the MSBs, and mapping of the two data streams onto either QPSK, 16-QAM, or 64-QAM constellations. These processes were described previously in Section 6.4.

6.5.2   Common Operating Characteristics of DVB-C Systems

Some of the more important parameters for DVB-C based QAM systems from a network operator's perspective are given in Table 6.4 below. These include the spectral efficiency (bits/Hz), the useful bit rate or payload (Mbps), and the required carrier to noise ratio (C/N) in decibels for a standard (8 MHz) channel [9]. As can be seen the highest data capacity is

Table 6.4
DVB-C Characteristics

Modulation (= 0.15)	16-QAM	32-QAM	64-QAM
Spectral efficiency (bits/Hz)	4	5	6
Useful bit rate (Mbps)	25.2	31.9	38.1
C/N in 8 MHz channel (dB)	19.7	23.3	25.7
Error correction	RS(204,188)	RS(204,188)	RS(204,188)

available with high order QAM systems (64-QAM), but these suffer from the highest C/N ratio requirement for error free operation.

6.6 Vestigial Sideband Modulation

The digital television standard prepared by the Advanced television systems committee (ATSC) technology group on distribution (T3), was approved as the United States advanced television system in 1995. The system employs MPEG-2 coding for video and data compression and uses the AC-3 system for digital audio compression. The ATSC digital television standard is described in document A/53 [3].

6.6.1 System Overview

The system can be used to transmit high quality video, audio, and data over a single 6 MHz broadcast channel. It was designed to deliver approximately 19.28 Mbps payload in a standard 6 MHz terrestrial broadcast channel using 8-level vestigial sideband modulation (8 VSB). It was also designed to be used in a cable environment using 16-level vestigial sideband modulation (16 VSB), and in this mode of operation it can deliver approximately 38.57 Mbits/s payload in a 6 MHz channel. Vestigial sideband modulation is a suppressed single carrier modulation scheme where most of the lower sideband is removed from the signal spectrum prior to transmission (hence, the name vestigial meaning "a trace of something formerly present"). It is generally regarded as a very robust system due to the presence of a pilot signal that helps receivers to demodulate the signal in poor reception conditions.

6.6.2 VSB Modulator

Many of the functions within a VSB modulator are similar to those found in single carrier QAM modulators, and (as we will see in Chapter 7) multicarrier modulators such as the COFDM system. The system functional block diagram is illustrated in Figure 6.10.

6.6.2.1 Data Adaptation and Outer Coding

At the input to the modulator the data is randomized to distribute the signal equally throughout the final RF channel bandwidth. This is achieved using a pseudo random binary sequence (PRBS) generator and XOR (exclusive or) circuitry. A more detailed description of the usage of a PRBS generator will be given in Sections 9.2.4 and 9.4. Reed Solomon (RS) coding is applied to each 188-byte MPEG-2 TS packet. The MPEG-2 TS packet contains 187 data bytes, which are error protected and randomized, and a sync byte, which is not adapted. The RS coding process adds 20 parity bytes for error protection to the 187 data bytes to give a total of 207 bytes. As a result it can correct up to 10 random errors in any packet, and the coding is therefore termed RS$(207,187, t = 10)$.

6.6.2.2 Interleaving and Trellis Coding

In a similar process to that described earlier for QAM systems, convolutional interleaving is used to protect the randomized data against a burst of errors together, which would overload the error correction capabilities of the receiver. An 8 VSB transmission also employs a rate $\frac{2}{3}$ trellis code system to give increased error protection.

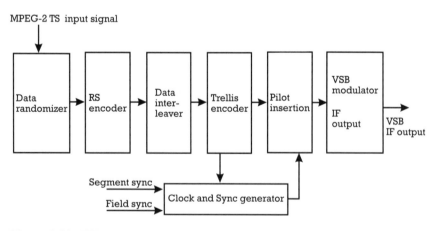

Figure 6.10 VSB modulator functional diagram.

6.6.2.3 Pilot Addition to VSB Signal

A small in phase pilot signal is added to the data signal at the same frequency and in place of the suppressed carrier signal, 310 kHz from the lower band edge, and approximately 11.3 dB below the average power of the data signal.

6.6.2.4 VSB Modulator

The signalling symbols used in the digital VSB transmission system are emitted at a rate of 10.76 Msymbols/sec, which corresponds to a Nyquist frequency of 5.38 MHz with an overall excess bandwidth of approximately 0.62 MHz. The total occupied bandwidth is divided into two identical parts, the upper band edge and the lower band edge, both with a linear phase raised cosine Nyquist filter response [10]. The system response is flat across the entire channel, except for the transition regions at each end of the channel [11]. The suppressed carrier and the pilot signal are located 0.31 MHz from the lower band edge as shown approximately in the spectrum plot of Figure 6.11.

6.7 Single Carrier and Multicarrier Modulation for Terrestrial Broadcasting

Variants of single carrier QAM, QPSK, and VSB systems can be used in cable, satellite, and MMDS transmission channels. As we have seen VSB modulation has also been specified for use in the United States as a

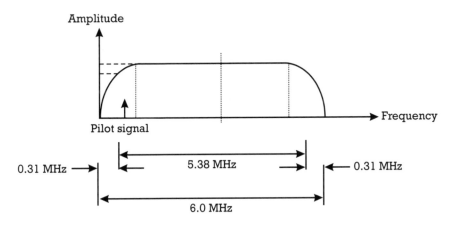

Figure 6.11 VSB spectrum.

terrestrial transmission method. However in Europe and elsewhere a different approach was adopted. Within the European DVB project a terrestrial commercial module was established to identify the requirements of a digital terrestrial system. The main findings of the group were that the terrestrial system should have the following features [12]:

1. The system should support the operation of single frequency networks (see Chapter 10).

2. The system should allow the operation of single frequency relays (gap fillers).

These requirements meant that only a multicarrier method of modulation would be suitable for the DVB-T (terrestrial) system. We will study this technology next in Chapter 7, and the topic of single frequency networks will be dealt with later in Chapter 10.

References

[1]   EBU/ETSI JTC, "Digital Broadcasting Systems for Television, Sound and Data Services; Framing Structure, Channel Coding and Modulation for 11/12 GHz Satellite Services," ETS 300 421, December, 1994.

[2]   EBU/ETSI JTC, "Digital Broadcasting Systems for Television, Sound and Data Services; Framing Structure, Channel Coding and Modulation for MMDS Below 10 GHz," ETS 300 749, January, 1996.

[3]   ATSC Standard A/53, Digital Television Standard, 1995.

[4]   de Bruin, R., and J. Smits, *Digital Video Broadcasting, Technology, Standards, and Regulations*, Norwood, MA: Artech House, 1999.

[5]   Kennedy, G., *Electronic Communication Systems*, McGraw-Hill, 1985.

[6]   Rogers, P. V., "Sound Broadcasting, AM to FM to DAB," *Proc. Callan Memorial Lecture, Institution of Engineers of Ireland*, February 10, 1994.

[7]   EBU/ETSI JTC, Digital Broadcasting Systems for Television, Sound and Data Services; Framing Structure, Channel Coding and Modulation for CATV Cable Systems, ETS 300 429, December, 1994.

[8]   "Digital television transmission," *ITS Technotes*, Vol. V, No. 3, December, 1992.

[9]   Cominetti, M., "Channel Coding and Modulation for Satellite and Cable," *EBU Digital TV Forum*, Prague, September 25–26, 1995.

[10] Eilers, C., "The In-band Characteristics of the Vestigial Side Band Emitted Signal for ATV Digital Terrestrial Broadcasting," *IEEE Trans. On Broadcasting*, Vol. 42, No. 4, December, 1996, pp. 298–304.

[11] Gumm, L., "Measurement of 8-VSB DTV Transmitter Emissions," *IEEE Trans. On Broadcasting*, Vol. 45, No. 2, June, 1999, pp. 234–242.

[12] DVB Document TM 1619, "Why DVB selected OFDM instead of single carrier modulation," February, 1996.

Coded Orthogonal Frequency Division Multiplexing

7.1   Introduction

Normal radio frequency (RF) transmission methods involve modulating data signals onto a single radio frequency carrier signal. In Chapter 6, all of the different digital modulation schemes used a single carrier to carry the data. Frequency Division Multiplexing (FDM) is a technique that is used in many applications to transmit simultaneously, many different data signals on different carrier signals. Simple telephony transmission systems use FDM, and in a similar manner multiple analog TV channels are broadcast simultaneously on different radio frequency channels. In all of these systems guard bands are employed between the RF signals to facilitate reception of the different signals on different frequencies. The reason that guard bands are used is because practical filters in receivers cannot select apart closely spaced carriers, and cross talk (interference between the carriers) would result. However, it is also possible to divide a single data signal into segments, such that each segment is at a much lower data rate than the original, and use each segment to modulate a single carrier from a multiplex or ensemble of radio frequency carriers.

Coded orthogonal frequency division multiplexing (COFDM) is a multicarrier modulation method that has been the subject of wide interest over many years, and first became attractive for transmission with the usage of the discrete Fourier Transform [1] and more recently due to implementations using Fast Fourier Transform (FFT) related techniques. It is very robust against multipath reception and is useful for channels that present linear distortions. It is the modulation that was chosen by the Digital Video Broadcasting (DVB) project (see Chapter 1) for development as a European standard for digital terrestrial television, and is known as DVB-T. It has been adopted by ETSI as a terrestrial standard (EN 300 744), and the reader is advised to refer to the standard for a complete description. However, the main features of COFDM and the DVB-T implementation are covered in this chapter. It is also used in digital audio broadcasting (DAB) as we shall see later in Chapter 13.

COFDM as mentioned above uses many thousands of separate carriers to convey the data signal, dividing the data between each carrier by a process of time division demultiplexing the data onto each carrier. The data signal is then modulated onto these closely spaced carriers using standard QPSK and QAM modulations, as discussed in Chapter 6. By selection of the right type of carrier signal it is possible to form an ensemble of carriers that are very tightly packed together and do not have need for a frequency guard band between them! This is a key attribute of COFDM, that is, the carriers are tightly packed into an RF channel without guard bands and yet each individual carrier can be received from within the ensemble.

Later in this chapter we will see that depending on the type of inner modulation chosen from the DVB-T modes of operation it is possible to achieve payload or useful bit rates between approximately 4 Mbps and 31 Mbps. This is a very attractive and flexible network design feature for the broadcaster. As we will see later in Chapter 10 and Chapter 13, COFDM technology also allows the possibility of establishment of a Single Frequency Network (SFN) for DTV signals, and also for digital audio broadcasting (DAB) signals. An SFN is a network where every transmitter broadcasts on the same RF channel, liberating valuable spectrum for other usage.

7.2   COFDM: A Qualitative Description

As mentioned in the introduction, the COFDM signal is comprised of an ensemble of closely spaced carriers generated by an Inverse Fast Fourier Transform (IFFT) technique within the modulator. The receiver uses a Fast Fourier Transform (FFT) in the demodulation process. The information data carried by the system is divided between the many carriers such that each carrier is modulated with a very low bit rate signal. Thus the loss of a carrier at the receiver has a minimal impact on the overall information transport, and error correction techniques can be used to mitigate against information loss despite the loss of carriers.

7.2.1   Frequency Division Multiplexing

The term, "frequency division multiplex," in the title COFDM arises from the fact that there are many thousands of closely spaced carriers generated by a COFDM modulator within the channel ensemble. These carriers are spaced at a fixed frequency separation from one another, and this frequency spacing is directly related to the useful data rate (termed active symbol period, Tu).

By dividing the data among a large number of closely spaced carriers, it is easy to ensure that each carrier contains only a small fraction of the overall data signal. This reduces the data rate of each carrier for a given modulation system (e.g., 64-QAM or QPSK) and lengthens the symbol period on each carrier. This means that if there is intersymbol interference, it will affect a smaller percentage of each symbol as the number of carriers and hence the symbol period increases.

7.2.2   Coded

The term, "coded," in the title is a reference to the channel coding employed by the COFDM modulation scheme to combat selective carrier fading. The distribution of the data signal over many carriers means that selective carrier fading will cause some bits to be received in error while some will be received correctly. By using an error correcting code which adds extra data bits at the transmitter, it is possible to correct many or all the bits that were incorrectly received. The information carried by one of the degraded carriers is corrected because other information, which is related to it by using this error protection code, is carried in another part of the carrier ensemble and received without error. COFDM typically uses a viterbi convolutional encoding system as the channel error protection code.

Carrier fading (which is common to all RF transmissions) would degrade the bit error rate (BER) performance of the system if channel coding were not employed. In fact the system would not work at all if channel coding were not used to distribute the data signal randomly among the carriers at different time instants. However, by employing a powerful viterbi code it results in making the system very robust in almost all fading conditions.

7.2.3 Orthogonality

The "orthogonality" condition mentioned in the COFDM modulation technique implies a definite and fixed relationship between all the carriers in the ensemble. It is a system requirement that the carriers are arranged such that the sidebands of each carrier overlap and can be received without intercarrier interference, hence the carriers must be orthogonal [2].

Orthogonality means that each carrier is positioned such that it occurs at the zero energy frequency point of all the other carriers. The $\sin(x)/x$ function has this property, and is used as a carrier in the COFDM modem, refer to the spectral diagram representation shown in Figure 7.1. Note that the side lobes of this carrier extend in both frequency directions and tend to zero as they move away from the main lobe. To ensure the orthogonality condition the frequency spacing of the carriers is fixed as the inverse of the active symbol period of the OFDM carriers. Then during the active symbol period (see below) every carrier is orthogonal to every other carrier. If then the time domain signal of any carrier, n, is multiplied by the time domain signal of another carrier m ($n \neq m$), and the product integrated over the active symbol period, the result is zero [3]. In practice this means that carriers never cause interference to each other.

7.2.3.1 Mathematical Meaning of Orthogonality

There are other ways of ensuring that signals are separated in time from each other (i.e., time division multiple access), or in frequency (i.e., frequency division multiple access). These are relatively simple ways of ensuring that signals do not interfere with each other. However, as we saw earlier, orthogonality is a commonly used technique to ensure that two signals occupying the same bandwidth can be separated by a receiver (see Section 6.4.1). Mathematically any two waveforms x and y are said to be orthogonal if their cross correlation $R_{xy}(0)$ over time interval T is zero, where

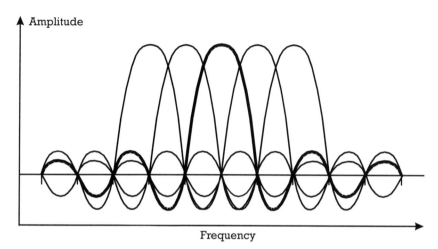

Figure 7.1 COFDM carriers.

$$R_{xy}(0) = \int_0^T x(t)\,y(t)\,dt \tag{7.1}$$

7.3 Symbol Period, Intervals, and Carrier Spacing

A symbol can be thought of as a period of time where a modulation state is applied to one or more of the carriers. It can consist of an active symbol period and a guard interval period. With COFDM, data is divided into active symbols (Tu), where an active symbol can comprise a modulation state applied to a particular carrier for a particular period of time (it can also represent many or all the carriers at a particular time instant) [4].

There is also defined a guard interval referred to as Δ (or sometimes as Tg), which is a period of time within each total symbol period when no new data is modulated onto the carriers. This guard interval is used to allow reception in a multipath environment, where time delayed signals are added to form a composite receive signal. During this guard interval any time delayed signal that is received will add constructively to the main signal without causing interference.

The summation of the active and guard interval gives the total symbol duration (Ts), as shown in Tables 7.1 and 7.2. This terminology causes much confusion and should be noted before reading further (see Figure 7.2). Note that using the guard interval in a symbol increases the duration of the total symbol and because the same information is broadcast

Table 7.1
Durations of Different Symbol Parts (*Source:* The present figures are the property of
ETSI and EBU. Further reproduction or use is prohibited. The original version of
the ETSI/EBU EN 300 744 can be obtained from the Publication Office of ETSI.)

Mode	8k mode				2k mode			
Duration of useful symbol part T_U	$8\,192{*}T$ (T is the elementary time period, and is the inverse of the system clock rate 64/7 MHz) 896 ms				$2\,048{*}T$ 224 ms			
Duration of guard Interval Δ	$2{,}048{*}T$ $224\,\mu s$	$1{,}024{*}T$ $112\,\mu s$	$512{*}T$ $56\,\mu s$	$256{*}T$ $28\,\mu s$	$512{*}T$ $56\,\mu s$	$256{*}T$ $28\,\mu s$	$128{*}T$ $14\,\mu s$	$64{*}T$ $7\,\mu s$
Guard interval Δ/T_U	$1/4$	$1/8$	$1/16$	$1/32$	$1/4$	$1/8$	$1/16$	$1/32$
Total symbol duration $T_s = \Delta + T_U$	$10{,}240{*}T$ $1{,}120\,\mu s$	$9{,}216{*}T$ $1{,}008\,\mu s$	$8{,}704{*}T$ $952\,\mu s$	$8{,}448{*}T$ $924\,\mu s$	$2{,}560{*}T$ $280\,\mu s$	$2{,}304{*}T$ $252\,\mu s$	$2{,}176{*}T$ $238\,\mu s$	$2{,}112{*}T$ $231\,\mu s$

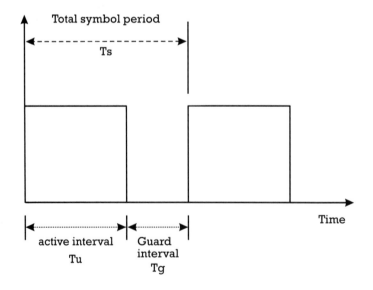

Figure 7.2   Total symbol period.

during the guard interval, it reduces the capacity of the symbol. As can be seen in Table 7.3, the guard interval duration, for a fixed modulation state and inner code rate, is increased at the expense of system capacity, however, it increases the multipath performance of the system and the SFN transmitter separation distance (see Chapter 10).

$$Ts = \Delta + Tu \tag{7.2}$$

As mentioned above the carriers are spaced in frequency by the inverse of the active interval (Tu). It is a property of any pulse modulation that the zero energy points of the modulated carrier signal are inversely proportional to the modulating pulse duration and not to the repetition rate of the total pulse train. This means that the zero energy points of the carriers do not change if the guard interval is shortened or lengthened, as long as the active symbol part of the total symbol remains fixed.

7.3.1 Consequence of Varying Guard Interval

Varying the duration of the guard interval will affect the total symbol duration (as this is the summation of the fixed active and variable guard intervals). This variation in total symbol duration due to varying the guard interval affects the information data rate of the modulator. As the total symbol duration increases, the data rate of the modulator is decreased. Hence in practical applications the guard interval must be traded carefully against the required data rate of the channel. Guard interval duration will affect the performance of the modem in the presence of multipath echo, and in single frequency networks set a limit to the maximum transmitter separation possible. Hence guard intervals are rarely set to the maximum possible duration as economic factors balance against technical performance. However, we will see that by selection of the 8k system, very long guard intervals are possible with even the shortest guard interval fraction, and this results in excellent multipath performance and high channel data capacity.

7.4 Time Domain Signal and Transmission Consequences

As can be seen from Figure 7.3 below the COFDM signal in the time domain exhibits a high peak to average power ratio (up to 12 dB) which

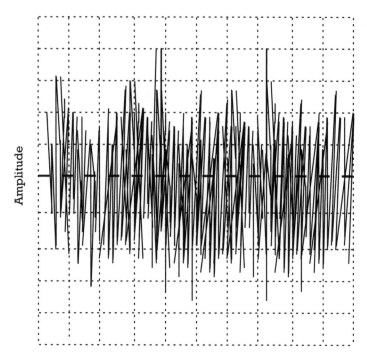

Figure 7.3   COFDM time domain signal.

makes it susceptible to non-linear distortion in transmission, as the signal peaks occasionally thrust the power amplifiers into saturation [5]. When this happens the transmitter will generate harmonics that will cause out of channel unwanted emissions, or interference. A practical consequence of this is that transmitter amplifiers must be operated in such a manner as to allow for these signal peaks and prevent amplifier saturation. This means that the transmitter output power back-off (OBO) must be adjusted to obtain the minimum required bit error rate (BER) perform-ance and also minimize adjacent channel interference. However, backing-off the power of a transmitter reduces it's efficiency and results in larger transmitter footprints for the rated output power of the device. However, it is expected that the footprints of digital transmitters will reduce with new cooling arrangements and digital precorrection. (See Chapter 8 for more information on transmitter architectures.)

As mentioned above, highly linear amplifiers are needed for digital transmitters and class AB operation is typical for solid state transmitters. While a 10 dB OBO is quoted for COFDM transmissions relative to analog transmissions to achieve similar coverage, most transmitter manufacturers recommend only a 6–7 dB reduction in transmitter output power when replacing an analog transmitter with a digital transmitter. In practical terms and as a general comparison, a 2 kW (rms) COFDM digital transmitter is presently comparable in the number of amplifiers used with a 10 kW (peak sync) analog transmitter. This is in order to achieve the required linearity. As we will see in Chapter 8, an output channel filter is also incorporated into many digital terrestrial transmitters to reduce the spurious emissions into adjacent channels.

7.5 The 2k and 8k Systems

In the DVB-T specification for COFDM implementation two different modes of operation are permitted, with different numbers of carriers based on either a 2,048 point Fast Fourier Transform (FFT), and known as the "2k" system, or based on an 8,192 point FFT, and known as the "8k" system.

The FFT size is given as a power of 2, where 2 is the FFT size. In the 2k system, n the power is 11 resulting in a 2,048 point FFT, and for the 8k system n is 13, resulting in an 8,192 point FFT. The actual number of carriers associated to each mode (which must be less than the FFT size) is given in Table 7.2 below, and includes those carriers that are transmitted with reference information for signalling purposes, and known as pilot carriers. The reference information is used by the receiver and refers to parameters such as inner code rate, hierarchy, guard interval, modulation, and transmission mode. Other information obtained by the receiver

Table 7.2
Numerical Values for COFDM Parameters for the 8k and 2k Modes

Parameter	8k mode	2k mode
Number of carriers K	6,817	1,705
Value of carrier number K_{min}	0	0
Value of carrier number K_{max}	6,816	1,704
Duration T_U	896 μs	224 μs
Carrier spacing $1/T_U$	1,116 Hz	4,464 Hz
Spacing between carriers	7.61 MHz	7.61 MHz
K_{min} and K_{max} $(K-1)/T_U$	(8 MHz channel)	(8 MHz channel)

from the usage of these so-called "continual pilot carriers" includes channel state information, phase noise tracking, and synchronization.

There are 1,705 carriers generated in the 2k mode while there are 6,817 carriers generated in 8k mode. However, many of these carriers are used for signalling and are not available to carry data payload. The number of useful carriers, available for data transmission, is 1,512 in the 2k mode and 6,048 in the 8k mode.

A common misunderstanding is that by choosing between the 2k and the 8k systems within the COFDM signal, it will affect the available useful data rate for a broadcaster; however, it will not. The main consequence of the choice to a broadcaster is that any selected guard interval (Δ) has a time duration that is four times longer in the 8k system than the corresponding guard interval duration in the 2k system. So for demanding reception conditions the 8k system will offer a fourfold increase in echo performance over that of the 2k system.

Historically the 2k system was perceived by many broadcasters as the more likely to be adopted, because a 2k Fast Fourier Transform (FFT) would be cheaper and easier to implement. This gave it a commercial advantage. It is also regarded as having better performance in mobile reception conditions. However, the 8k system, which has longer guard intervals available without a reduction in data payload, is now becoming the most implemented variant of the COFDM standard. All 8k system receivers can decode the 2k signals, but not vice versa.

7.5.1 Data Payload of 2k and 8k Systems

The overall bit rate available for data transmission with COFDM systems is not dependent on the choice between a 2k or 8k system. It depends upon the modulation type used to map information onto each carrier, the guard interval duration selected, and the code rate selected (see Section 7.6), as can be seen from Table 7.3. The basic modulations available to choose from include QPSK, 16-QAM, and 64-QAM. QPSK will be the most resilient in low power budget applications but will lose valuable payload capacity. 64-QAM is the most common mode selected for broadcasting as it has the highest data capacity.

Note that by selection of appropriate modulation, guard interval, and code rate, the system designer has between 4.98 Mbps and 31.67 Mbps data payload capacity to choose from. Clearly QPSK modulation will only be used within COFDM where extremely rugged signal performance is required, due to the low data rate available. The most common system operating modes will use 64-QAM with a short guard interval and strong

Table 7.3

Useful Bit Rate (Mbps) for All Combinations of Guard Interval, Constellation, and Code Rate (for Nonhierarchical Systems) (*Source:* The present figures are the property of ETSI and EBU. Further reproduction or use is prohibited. The original version of the ETSI/EBU EN 300 744 can be obtained from the Publication Office of ETSI.)

Modulation	Code rate	Guard interval			
		1/4	1/8	1/16	1/32
QPSK	1/2	4,98	5,53	5,85	6,03
	2/3	6,64	7,37	7,81	8,04
	3/4	7,46	8,29	8,78	9,05
	5/6	8,29	9,22	9,76	10,05
	7/8	8,71	9,68	10,25	10,56
16-QAM	1/2	9,95	11,06	11,71	12,06
	2/3	13,27	14,75	15,61	16,09
	3/4	14,93	16,59	17,56	18,10
	5/6	16,59	18,43	19,52	20,11
	7/8	17,42	19,35	20,49	21,11
64-QAM	1/2	14,93	16,59	17,56	18,10
	2/3	19,91	22,12	23,42	24,13
	3/4	22,39	24,88	26,35	27,14
	5/6	24,88	27,65	29,27	30,16
	7/8	26,13	29,03	30,74	31,67

error correction. A typical choice will be 64-QAM, Guard Interval = $\frac{1}{32}$, and Code Rate = $\frac{2}{3}$, which gives a payload of 24.13 Mbit/second. The choice is left to the network operator to trade performance with data payload. In Table 7.4 below is shown a comparison of the transmission efficiency of the different modulations and code rates for a fixed Guard Interval (GI) of $\frac{1}{4}$. It is evident that most network operators and broadcasters will tend toward 64-QAM modulations due to the increased capacity of these systems over 16-QAM and QPSK.

By selection of a shorter guard interval it is possible to increase the transmission efficiency of the channel. As can be seen in Table 7.5 with a guard interval = $\frac{1}{32}$ the data capacity of a COFDM channel is much greater. Broadcasters are attracted to low guard interval modes within

Table 7.4
Transmission Efficiency in Bits/Hz for DVB-T Modes With GI = $\frac{1}{4}$

Code rate	QPSK	16-QAM	64-QAM
1/2	0.62	1.24	1.87
2/3	0.83	1.66	2.49
3/4	0.93	1.87	2.80
5/6	1.04	2.07	3.11
7/8	1.09	2.18	3.27

COFDM for this reason. However, in this case multipath performance will be reduced to gain data capacity. But by selection of the 8k system variant, all guard interval durations are increased by a factor of four in absolute time duration over those available in the 2k system. This is why the 8k system is so popular with network operators who need to use efficient broadcasting modulations and maintain good multipath performance.

The main difference between the 2k and the 8k system variants is in performance with echoes and in transmitter separation in single frequency networks. This is due to the different durations of the guard intervals available in both variants. As mentioned above, the 8k system supports guard intervals which are four times longer in duration than those of the 2k system. It can therefore tolerate reception of strong echoes that are four times longer than the maximum echo a 2k system can tolerate. By the same process it can be used in SFNs that have transmitters with four times larger separation than those used in 2k-based SFNs (as we will see in Chapter 10). As we have just seen, the 8k variant can

Table 7.5
Transmission Efficiency in Bits/Hz for DVB-T Modes With GI = 1/32

Code rate	QPSK	16-QAM	64-QAM
$\frac{1}{2}$	0.75	1.50	2.26
$\frac{2}{3}$	1.01	2.01	3.01
$\frac{3}{4}$	1.13	2.26	3.39
$\frac{5}{6}$	1.25	2.51	3.77
$\frac{7}{8}$	1.32	2.63	3.95

also allow reduction in guard interval fraction from $\frac{1}{4}$ to $\frac{1}{32}$, while maintaining adequate terrestrial multipath performance.

7.6 Inner Coding (Convolutional Codes)

Inner coding and modulation are among the parameters that the network designer has control over, and as such are more important for discussion than the outer coding (which is performed prior to inner coding) techniques. This is why they are treated first in this chapter.

As mentioned above, the network designer can choose between many different operating modes when using COFDM technology. Some of these choices are for the type of inner coding scheme used. A Rate Compatible Punctured Convolutional (RCPC) code has been defined for the DVB-T system [3, 6]. The inner coding is a convolutional code, which for each input bit produces two output bits. The coding rate in this case is $\frac{1}{2}$, as the coding rate is the number of information bits divided by the total number of bits.

Via a procedure known as puncturing, it is possible to throw away in a systematic way, some of the output bits of a convolutional encoder. This means that one input bit after puncturing no longer produces two output bits, but less than this, hence the coding rate is increased. There are five choices of inner code rates, ranging from $\frac{1}{2}$ code rate (unpunctured), to $\frac{2}{3}$, $\frac{3}{4}$, $\frac{5}{6}$, and $\frac{7}{8}$. This will allow the designer to select an appropriate level of error correction for a service. The highest error protection is afforded to code rate 1/2 as this has the highest redundancy. However it protects at a price, data rate is reduced when high error correction is used.

After this process the data is demultiplexed onto different parallel data streams in preparation for mapping onto the carriers. The data is processed in different ways depending on what inner modulation has been selected by the network designer.

It is fundamental to efficient decoding of signals in fading channels that the receiver knows in advance that some bits are reliable. A convolutional decoder which normally is a so-called Viterbi decoder, knows a priori where there are missing bits from the puncturing process and inserts zeros at these points. By a process known as "soft decision decoding" the Viterbi decoder can distinguish between reliable and unreliable bits [6].

7.7 Inner Modulation

There is often confusion about the inner modulations that COFDM supports. COFDM can use QAM or QPSK modulations to modulate each carrier separately, and the total resulting ensemble of carriers is the COFDM signal. QAM and QPSK are also used for single carrier systems such as those used for satellite and cable transmission (see Chapters 6 and 12). COFDM, however, is a multicarrier system that uses QAM modulations to map information onto each of many thousands of carriers. It has the added multipath and SFN performance which basic QAM systems do not have.

There are three basic modulations available for usage with COFDM, these are QPSK (or 4-QAM), 16-QAM, and 64-QAM. As mentioned above every carrier is modulated with one of these modulation schemes selected by the network designer. QPSK will map 2 bits per symbol onto a carrier, while 16-QAM will map 4 bits per symbol onto a carrier, and 64-QAM will map 6 bits per symbol onto each carrier. The mapping process allows the formation of constellations in the phase space plane and uses Gray mapping. This means that adjacent constellation points differ by only one bit, so if the wrong constellation point is decoded by accident it should be close in value to the correct point.

7.8 Outer Coding

COFDM modulators accept as input signals an MPEG-2 transport stream. The MPEG-2 transport stream will be already organized into fixed length packets of data, and the packet length will be 188 bytes, of which 187 will contain data and 1 will be a sync byte (see Figure 7.4a).

This 187 byte packetized data stream must first be randomized before transmission to ensure that the COFDM signal has equal information dispersed throughout its entire bandwidth. This is achieved by passing it through a scrambler which uses a pseudo random binary sequence (PRBS) generator to ensure that the MPEG-2 signal has all its information data randomly distributed throughout the data stream. It also ensures that there are plenty of binary transitions, i.e., that there are no long strings of 1's and 0's. This will help to combat against bursts of errors while in transmission. The PRBS generator used has a period of 1,503 bytes and it can deal with 8 packets of data at a time (see Figure 7.4b).

Next a Reed-Solomon code RS (204,188,8) is applied to each randomized MPEG-2 packet. This code has a length of 204 bytes, 16 of which are parity bytes, and the remaining 188 bytes contain the randomized data and sync byte (see Figure 7.4c). The code can correct for 8 randomly errored bytes out of a 204-byte word. Finally the data packets are interleaved and passed to the inner coder and modulator.

7.9 Pilot Cells

The COFDM signal comprises an ensemble of separately modulated carriers, where the data carried by one carrier at a particular time instant is regarded as a cell. The complete ensemble at a particular time instant is regarded as a symbol, and 68 of these symbols make a frame.

A COFDM frame carries data on most of the carriers but some of the carriers are reserved for frame and time synchronization, channel estimation, and for transmission parameter signaling (TPS). These carriers or pilot cells are located throughout the carrier ensemble according to defined rules and are often transmitted at slightly higher power levels.

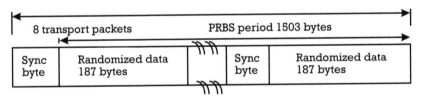

Sync byte	MPEG2 data, 187 bytes

[a] MPEG2 TS MUX packet

8 transport packets   PRBS period 1503 bytes

Sync byte	Randomized data 187 bytes		Sync byte	Randomized data 187 bytes

[b] Randomized packets

Sync byte	Randomized data 187 bytes	16 parity bytes

[c] Reed Solomon error protected packets RS(204,188,8)

Figure 7.4   MPEG-2 TS packet processing.

They help the receiver to establish contact and they pass on valuable information to the receiver about the type of modulation chosen, the guard interval duration to expect, and the code rate used. They can be seen on an oscilloscope when a constellation diagram of a COFDM based DVB-T signal is viewed.

7.10   Radio Frequency Spectrum Characteristics

As mentioned above the COFDM symbols constitute an ensemble of equally spaced orthogonal carriers. The amplitudes and phases of the data cell carriers are varying symbol by symbol according to the mapping process associated with the chosen modulation.

The power spectral density $P_k(f)$ of each carrier k, at frequency

$$\mathbf{f}_k = \mathbf{f}_c + \frac{k'}{\mathbf{T}_u}$$

$$\mathbf{k}' = \mathbf{k} - (\mathbf{K}_{max} + \mathbf{K}_{min})/2; (\mathbf{K}_{min} \leq \mathbf{k} \leq \mathbf{K}_{max})$$

is defined by the following expression:

$$P_k(f) = \left[\frac{\sin\pi \cdot (f - f_k) \cdot T_s}{\pi \cdot (f - f_k) \cdot T_s} \right]^2 \tag{7.3}$$

where $\mathbf{f}_c$ is the central frequency of the RF signal, T_s is the total symbol duration, $\mathbf{k}$ is the carrier index relative to the center frequency, and $\mathbf{K}_{max}$ and $\mathbf{K}_{min}$ are given in Table 7.2.

The overall power spectral density of the modulated data cell carriers is the sum of the power spectral densities of all of these carriers. A theoretical COFDM transmission signal spectrum is illustrated in Figure 7.5. Because the COFDM symbol duration is larger than the inverse of the carrier spacing, the main lobe of the power spectral density of each carrier is narrower than twice the carrier spacing. It should be noted that practical COFDM modulators do not produce signal spectra that show such a marked contrast in signal spectrum between the 2k and 8k variants.

7.10.1 RF Channel Bandwidth

The COFDM signal bandwidth can be tailored for different channel bandwidths. The typical application is for an 8 MHz wide UHF channel and this is easily modified for 7 MHz channels by scaling all parameters in accordance to the change in system clock rate from 64/7 MHz to 8.0

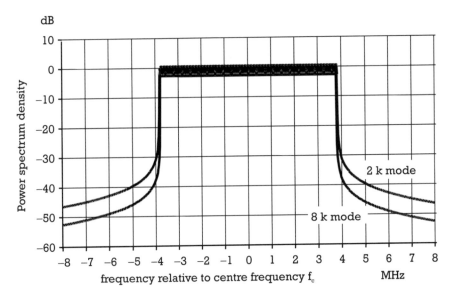

Figure 7.5   Theoretical COFDM transmission spectrum.

MHz. This book uses 8 MHz as a standard channel bandwidth. The nominal bandwidth in an 8 MHz channel is approximately 7.608254 MHz for the 8k mode and 7.611607 MHz for the 2k mode [6]. When reducing the channel bandwidth down to 7 MHz a corresponding reduction of 7/8 can be expected in system data capacity. The new signal bandwidth will be approximately 6.6 MHz for a 7 MHz channel bandwidth. The COFDM signal is normally centred co-incidentally with the centre of the assigned Radio Frequency channel. This leaves approximately 0.2 MHz on either side of the COFDM signal as a guard band and relaxes somewhat the required roll-off of transmitter output channel filters.

7.11   Hierarchical Transmission

This is a modulation technique that allows the partition of a broadcast channel into two layers, each with different levels of service. One layer of service carries a low bit rate MPEG-2 signal that is highly error protected and uses QPSK as the inner modulation. The other layer of service carries a less rugged signal at a higher bit rate using either 16-QAM or 64-QAM as the modulation.

A unique feature of DVB-T, the terrestrial digital standard, is this hierarchical transmission which is not yet incorporated in other transmission standards. With the COFDM-based terrestrial standard it is presently possible to incorporate hierarchical transmission, where an incoming multiplex is split into two separate streams, known as a low priority (LP) and a high priority (HP) stream with different channel coding and modulation applied to each stream. A COFDM modulator functional block diagram is shown in Figure 7.6 along with a notional MPEG-2 TS splitter. In practical modulator realizations the HP and LP streams are input to the modulator on different physical input connections.

This system could allow the broadcaster to separate different programs within a stream into two layers and to afford to each a different level of ruggedness due to different selections of coding and modulation. It is possible to use nonuniform constellations (multiresolution QAM) to achieve the desired coverage for each of the two layers [7]. Ordinary QAM has equal separation between the various amplitude levels used in the quadrature modulating waveforms. Nonuniform QAM has significantly different separation between levels in the I and Q quadrants. With these nonuniform constellations it is possible to embed a QPSK signal into either a nonuniform 16-QAM signal as shown in Figure 7.7, or a

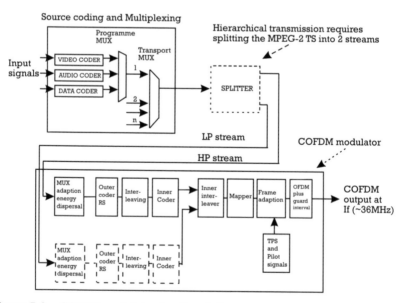

Figure 7.6   COFDM modulator functional diagram.

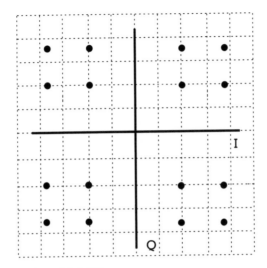

Figure 7.7 Nonuniform 16-QAM.

nonuniform 64-QAM signal as shown in Figure 7.8. These low priority data stream constellations are often referred to as "clouds" because the complete quadrant defines either a range of symbols when the carrier to noise (C/N) conditions are good, or shrinks to define one QPSK symbol

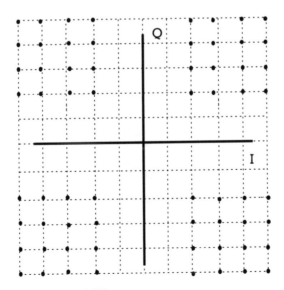

Figure 7.8 Nonuniform 64-QAM.

when C/N conditions are bad [8]. Note that the higher order QAM symbols in any quadrant of the I/Q plane represent the corresponding QPSK quadrant symbol (see Figure 7.9). The increased ruggedness afforded to one channel in the multiplex increases the coverage of that channel but reduces the total multiplex bit rate slightly.

As an illustration of the coverage benefits of hierarchical transmission, the above mentioned QPSK signal can be embedded within a non-uniform 64-QAM signal and both transmitted together (see Table 7.6). The more robust QPSK signal at a bit rate of 4.98 Mbps, code rate $\frac{1}{2}$, guard interval $\frac{1}{4}$, would require a carrier to noise ratio of only 7.1 dB (in a Ricean channel) for reception, while the 64-QAM signal transmitted with it at a higher bit rate of 16.59 Mbps, code rate $\frac{2}{3}$, and similar guard interval duration would require a much higher C/N ratio of 22.7 dB within the same channel (see Figure 7.10). It then allows the two different MPEG-2 TS to be broadcast with the one transmitter, one as a QPSK signal and the other as a 64-QAM signal.

The total bit rate for this hierarchical mode of operation is reduced to 21.57 Mbps (4.98 Mbps + 16.59 Mbps), whereas if nonhierarchical operations were used the transmission rate would be 24.88 Mbps with a 64-QAM signal using similar code rate and guard interval [9]. As a

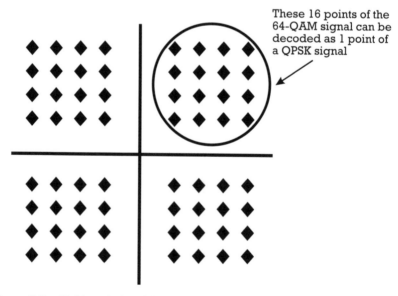

These 16 points of the 64-QAM signal can be decoded as 1 point of a QPSK signal

Figure 7.9   Multiresolution QAM principal.

Table 7.6
Required C/N for Hierarchical Transmission to Achieve a BER = 2 10⁻⁴ After Viterbi Decoder (Source: The present figures are the property of ETSI and EBU. Further reproduction or use is prohibited. The original version of the ETSI/EBU EN 300 744 can be obtained from the Publication Office of ETSI.)

Modu-lation	Code Rate	α	Required C/N for BER= 2 × 10⁻⁴ after Viterbi QEF after Reed-Solomon			Bit rate (Mbps)			
			Gaussian Channel	Ricean Channel (F)	Rayleigh Channel (P1)	$D/T_v =$ 1/4	$D/T_v =$ 1/8	$D/T_v =$ 1/16	$D/T_v =$ 1/32
QPSK	1/2		8,9	9,5	11,4	4,98	5,53	5,85	6,03
	2/3		12,1	12,7	14,8	6,64	7,37	7,81	8,04
	3/4		13,7	14,3	17,5	7,46	8,29	8,78	9,05
in		1	+						
	1/2		14,6	14,9	16,4	9,95	11,06	11,71	12,06
uniform	2/3		16,9	17,6	19,4	13,27	14,75	15,61	16,09
64-QAM	3/4		18,6	19,1	22,2	14,93	16,59	17,56	18,10
	5/6		20,1	20,8	25,8	16,59	18,43	19,52	20,11
	7/8		21,1	22,2	27,6	17,42	19,35	20,49	21,11
QPSK	1/2		6,5	7,1	8,7	4,98	5,53	5,85	6,03
	2/3		9,0	9,9	11,7	6,64	7,37	7,81	8,04
	3/4		10,8	11,5	14,5	7,46	8,29	8,78	9,05
in		2	+						
	1/2		16,3	16,7	18,2	9,95	11,06	11,71	12,06
non-	2/3		18,9	19,5	21,7	13,27	14,75	15,61	16,09
uniform	3/4		21,0	21,6	24,5	14,93	16,59	17,56	18,10
64-QAM	5/6		21,9	22,7	27,3	16,59	18,43	19,52	20,11
	7/8		22,9	23,8	29,6	17,42	19,35	20,49	21,11

generalized rule of thumb, hierarchical transmission reduces the data payload by approximately 10% compared to nonhierarchical 64-QAM modulation. It also requires nearly 3 dB extra carrier to noise ratio margin for the 64-QAM LP stream, compared to that required for nonhierarchical transmission.

7.11.1 Possible Uses of Hierarchical Transmission
This transmission system is very new to the broadcasting industry and has yet to be deployed in commercial applications. It is predicted that the

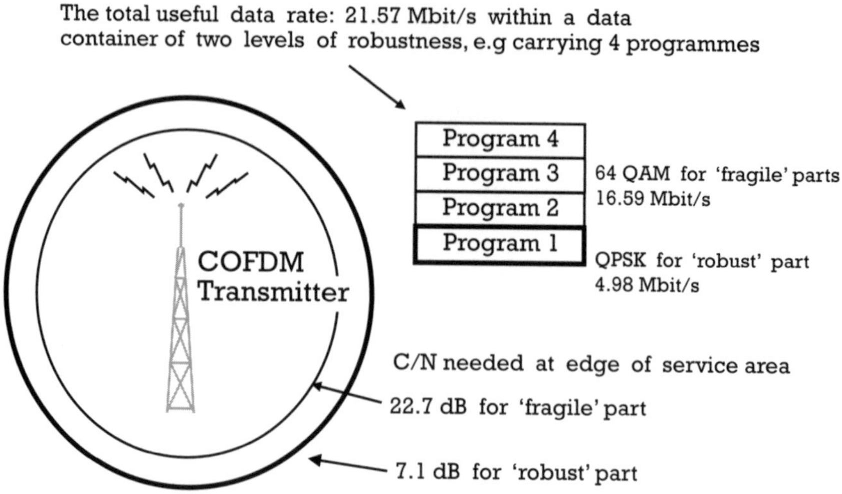
The total useful data rate: 21.57 Mbit/s within a data container of two levels of robustness, e.g carrying 4 programmes
Program 4
Program 3
Program 2
Program 1
64 QAM for 'fragile' parts
16.59 Mbit/s
QPSK for 'robust' part
4.98 Mbit/s
COFDM
Transmitter
C/N needed at edge of service area
22.7 dB for 'fragile' part
7.1 dB for 'robust' part

Figure 7.10 Hierarchical transmission example.

hierarchical transmission technique could be used to simulcast a high bit rate, high definition television (HDTV) MPEG-2 transport stream together with a lower bit rate, standard definition television (SDTV) MPEG-2 transport stream, within the same COFDM signal. The HDTV signal (LP) would be available only to those consumers in the well covered areas of the transmitter main service area, while the SDTV signal (HP) would propagate further and be available for reception in other areas beyond the main service area.

Another possible usage would be to allow mobile reception of a low bit rate QPSK modulated COFDM signal in cars and trains, while also supporting higher bit rate broadcasts to fixed receivers using 64-QAM modulation within COFDM. Hence mobile reception could be supported at low payload bit rates, in conjunction with fixed reception of higher payloads.

7.12 Practical DVB-T Modulators

There are now COFDM modulators available from many different vendors that support all the DVB-T modes. Manufacturers are bringing newer and newer versions of DVB-T modulators to the market with great haste to take advantage of growing markets. Currently available equipment tends to accept as input signal an MPEG-2 TS in the SPI and/or the ASI electrical format (see Chapter 5). The modulator may also accept a timing reference and a frequency reference signal for single frequency network (SFN) operation (we will see this later in Chapter 10). Few modulator implementations are, as of now, capable of two level hierarchical modulation due to the lack of demand for this feature. However, those that do support the scheme have a high priority and a low priority MPEG-2 TS input.

Typically a modulator will also be capable of generating a Pseudo Random Binary Sequence (PRBS) for Bit Error Rate (BER) measurements. The BER can then be measured at many different reference points within a DVB-T receiver but some points are more useful then others for estimating BER performance and Carrier-to-Noise ratio margin (see Chapter 9). Some modulators have advanced features such as the removal of a number of adjacent carriers within the COFDM ensemble to allow in-band distortion measurements. Other features include some level of signal precorrection to counteract transmitter distortion of the COFDM signal during amplification.

The COFDM modulator output signal is generally an intermediary frequency (If) signal which is centered at approximately 36 MHz. This is the input to a transmitter exciter (see Figure 7.6). However many modulators are now equipped with digital I and Q signal interfaces as well as the If interface. The advantage of using the I and Q interface signals is that it allows the usage of a completely digital precorrection process on the COFDM signal, to counteract the distortion produced in the transmitter power amplification stages. This precorrection can be made into an automatic self-alignment process for DTV modulators and transmitters.

7.12.1 Use of FFT in COFDM

There is great interest in how the thousands of carriers are actually generated within a COFDM modulator. The main reason that the COFDM technique has taken so long to be developed is practical. A hardware solution using banks of oscillators in parallel was never practical to realize except in a laboratory. Now using the Fast Fourier Transform technique it is possible to define the signal in the frequency domain, in software, and to generate the signal using the inverse FFT algorithms that are common to digital signal processing (DSP) chips. At the transmitter, the signal is therefore defined in software in the frequency domain. It is a sampled digital signal, and it is defined such that the discrete Fourier spectrum exists only at discrete frequencies. Each COFDM carrier corresponds to one element of this discrete Fourier spectrum. To enable the signal to be generated using an inverse FFT, it is necessary to use a power of 2 in the calculation. This is why the commonly used 2k, or 2,000 point inverse FFT, is used in terminology. It is a similar procedure for the 8k, or 8,000 point system.

References

[1]   Weinstein, S. B., and P. M. Ebert, "Data Transmission by Frequency Division Multiplexing Using the Discrete Fourier Transform," *IEEE Transactions on Communication Technology*, COM-19, No. 5, October 1971, pp. 628–634.

[2]   Shelswell, P., "The COFDM Modulation System: The Heart of Digital Audio Broadcasting," *Electronics & Communication Engineering Journal*, June 1995, pp. 127–136.

[3]   Stare, E., "Channel Coding and Modulation for Digital Terrestrial TV," *EBU Digital TV Forum*, Prague, Sept. 25–26, 1995.

[4]   Rogers, P. V., "Sound Broadcasting, AM to FM to DAB," *Proc. Callan Memorial Lecture, Institution of Engineers of Ireland*, February 10, 1994.

[5]   Qun Shi, "OFDM in Bandpass Nonlinearity," *IEEE Trans. Consumer Electron*, Vol. 42, No. 3, August 1996.

[6]   DVB Document A012, "Framing Structure, Channel Coding and Modulation for Digital Terrestrial Television," EN 300744, V1.1.2 (1997-08).

[7]   Moller, L., "Digital Terrestrial Television, The 8k System," *EBU Technical Review*, Winter 1995, pp. 40–50.

[8]   Reimers, U., "DVB-T: the COFDM-Based System for Terrestrial Television," *Electronics & Communication Engineering Journal*, Vol. 9, No. 1, February 1997, pp. 28–32.

[9]   O'Leary, S., "Hierarchical Transmission and COFDM Systems," *IEEE Trans. On Broadcasting*, Vol. 43, No. 2, June 1997, pp. 166–171.

Contents

Terrestrial Transmission of DTV Signals

8.1   Introduction

This chapter seeks to describe some aspects of the transmission of digital television (DTV) signals. It will describe the advantages of using digital signals for transmission to the home compared with the transmission of analog signals as currently predominates in broadcasting. The relative advantages of using the UHF and VHF spectrum for transmission purposes, and state of the art digital transmitters will be described, as well as high power channel combiners and channel filters. It will review analog terrestrial channel assignment and the reasoning behind the channel grouping methods used for the PAL I systems. The requirements for power handling capability and the frequency response needed will be discussed. Other important transmitter performance parameters such as linearity, precorrection, and output back off

(OBO) will be discussed. Finally power consumption and transmitter cooling methodologies will be addressed.

8.1.1 Digital and Analog Transmitted Picture Quality

The transmission of television signals on all media (terrestrial, cable, satellite, and MMDS), whether in analog or digital form requires the usage of some encoding scheme. Broadcasters have used the current analog (PAL and NTSC) composite coding schemes for decades, and these transmit color information on subcarriers, which often causes the smearing of colors on detailed picture elements. However by digitally encoding an analog signal while still in component format this type of degradation can be avoided. Broadcasting involves the transmission of the information signal in either analog or digital format to the consumer on a specific transmission medium (cable, terrestrial, and so forth) and this, as we will see, introduces other signal impairments. However, digital signals can use powerful error correction schemes, as was discussed in Chapter 6 and Chapter 7, to remove any errors introduced in transmission. This means that the digital signal received by a consumer can be of the same quality as that of the originating encoder. This was not possible with analog transmission methods as noise was always added in transmission, and it could not be corrected.

8.1.2 Analog Transmission Impairments

Analog transmission on any medium is often susceptible to severe picture ghosting due to many different causes. These can include simple occurrences such as signal reflections from buildings and obstructions, which add to the main receive signal at a television receiver, and have a time delay that is due to the different paths taken by the reflected signals. It can also be due to terrain effects such as reflections from mountains and hills. Other classical signal degradations include tidal effects, and these occur when signals have to traverse large amounts of open seawater (see Chapter 14). As the tide rises and falls during the day, the length of the reflected signal paths change due to the change in sea level, which results in slowly changing multipath reflection conditions. Other classical effects known as co-channel interference (CCI) are noted during periods of atmospheric high pressure, and are caused by the operation of different transmitters on the same assigned frequency. These multipath reflection and co-channel interference problems have long since been associated with analog broadcasting and have resulted in the international adoption of abstract protection rules by neighboring countries to minimize interference from co-channel sites. Despite these rigorous protection rules, analog television is sensitive to degradation while in transmission as the process always adds some noise to the signal, even in the best receiving conditions. These protection rules will be treated in more detail later on in Chapter 11.

8.1.3 Digital Transmission Advantages

The inherent advantage of digital transmission is that the signal can be decoded without the addition of the errors mentioned in Section 8.1.1. There will still be noise added to a digital signal in transmission, and signal reflections, and ghosts will also arrive at the input to a digital receiver with the same delays as before due to the same path differences. However, a digital receiver should be able to separate the wanted signal from the noise and ghosts to reconstruct the transmitted signal that left the studio. By decoding the signal into component format the digital signal will avoid the loss of clarity introduced by the analog (PAL/NTSC) encoding formats. As was discussed in the earlier chapters dealing with encoding and multiplexing, the broadcaster will be able to set the technical visual and aural quality of a program by allocating a bit rate to that program service in the studio with the MPEG-2 encoder. The resolution of the received picture will not be varied by the bit rate, but the picture will start to have artifacts if the bit rate is to low for the detail of the program content. Digital transmission means that picture quality will become independent of transmission artifacts, and at last the quality of the receive signal will be controlled more by the encoding and multiplexing system than by the channel encoding system and transmission path. Other advantages include spectrum efficiency, which means that more services can be accommodated in digital format within a radio frequency channel than in analog format (see Chapters 6 and 7). Digital format information can carry any signal type, including audio, video, and data. Analog format information cannot integrate these signals as easily.

8.2 Transmission Spectrum Issues

Broadcasters can use various segments of the electromagnetic spectrum for transmission of television and radio services. Each potential transmission medium has intrinsic propagation properties, which limit the extent of coverage that can be achieved with it, and there are always economic reasons that favor the usage of a particular medium. The various media include cable, terrestrial, satellite, and MMDS. As will be seen, historically terrestrial transmission was the first medium established. Chapter 13 will detail the developments in another terrestrial broadcasting scheme, MMDS. This section will detail some of the developments in terrestrial spectrum allocation and indicate the possible spectrum choices for DTV service provision.

8.2.1 Analog Spectrum Allocations and Frequency Bands

The British Broadcasting Corporation (BBC) initiated analog television transmissions on the medium waveband using a mechanical scanning system as far back as early 1929. In 1937 the United Kingdom adopted the 405-line monochrome system and used the frequency band known as band I (approximate frequency range 45–68 MHz) for transmission channels. However, this allowed only the usage of five radio frequency channels for television broadcasting. With the development of a second U.K. service in 1955 (known as Independent Television, or ITV) it became clear that band III (approximate frequency range 174–230 MHz) would be a better band to operate in as it would allow the development of a national two channel network in the United Kingdom. The 405-line system came to an end officially in 1985, when the United Kingdom moved to the present analog 625-line PAL system standard and moved to the UHF bands IV and V (470–860 MHz). This allows national coverage of four channels and limited coverage with a fifth channel at present.

In other parts of the world mixtures of band I, band III, and bands IV and V are used to accommodate terrestrial transmissions. There is a migration out of band I to allow other users access to this spectrum. Band I has excellent propagation characteristics, but is prone to suffer from co-channel interference at times of enhanced propagation. For example, until recently from time to time co-channel interference was noted, in the south of Ireland on signals received from (the last remaining) band I transmitter in Ireland, which originated from a co-channel transmitter in Spain. DTV will not be deployed in band I because the bandwidth is too narrow for many channels, and it is designated for other usages in the near future.

8.2.2 Television Planning

As we shall see later in Chapter 11, radio frequency (RF) spectrum is a scarce and valuable commodity. Once allocated to a particular service, a frequency or channel cannot be re-used within that service area for any other service. Frequency re-use is necessary to allow many service providers access to channels. As a result frequency re-use is not only required, it is a necessity for service provision and as a result very many co-channel broadcasts are made at any time on any assigned channel, as can be seen in Figure 10.1. Interference from co-channel (termed CCI, co-channel interference) transmissions is of major concern to all users of radio frequency channels as it impairs and often destroys completely the wanted signal. Terrestrial television signals can propagate long distances when climatic conditions are favorable, far beyond the planned service area of the mainstation transmitter, and hence cause interference in other service areas that use the same RF channel.

As a result of this problem in 1961, the Stockholm agreement was established, and this agreement proposed a set of frequency allotments for many countries around the world to allow the establishment of multi-channel terrestrial television networks. The Stockholm plan is a list of maximum transmitter Emitted Radiated Powers (ERPs) for particular mainstation sites using specified aerial heights. In general it allows for national coverage of a four-channel UHF terrestrial television network, which will have minimal interference with adjacent countries' transmitters.

8.2.3 UHF Propagation

It is well known that in the UHF band, transmissions are limited to a little more than line-of-sight propagation. The signals tend to diffract only slightly around terrain, unlike VHF transmissions, which experience greater diffraction around hills and mountains and therefore cover isolated valleys and hamlets to a greater extent. It is very easy to predict the transmitter service area at UHF, once the terrain data is available together with the transmitter location, ERP, and aerial height.

8.2.3.1 Transmitter Service Area and Coverage

When planning a transmitter network great care is taken when selecting sites to meet objectives such as the maximization of population coverage, the minimization of CCI, and other issues such as environmental impact, ease of access, and cost.

By locating a transmitter mainstation close to centers of population and by using aerial beam tilt on the transmit antenna system [1] to direct the main lobe of radiated power below the horizon, it can be arranged that little radiation is transmitted beyond the service area. The antenna gain will tend to concentrate the transmitter radiation into a beam, which will ensure that adequate field strength is achieved for reception of services within the service area of that transmitter. The gain process is similar to that of placing a magnifying glass into strong sunlight, the radiation from the antenna is concentrated into a beam, which can be directed toward the ground within the service area.

The service area of a transmitter is limited by among other things, the earth's curvature, so that for a transmitter broadcasting from an aerial at a height h (meters) above sea level, the edge of the service area will be located at a radial distance d (km) away, where

$$d = \sqrt{17h} \quad [km] \tag{8.1}$$

We shall see later in Section 12.2.1 that the radio horizon, or service area radius of a transmitter can be extended beyond this value if climatic propagation conditions are favorable.

Example 8.1

A 60m mast with a UHF panel located at the top of the mast is erected on a hill with a 450m elevation above sea level. What will be the radius of the service area, assuming a flat topography?

The antenna is positioned 60 + 450 = 510 meters above sea level. Then the expected service area horizon (d) will be

$$h = 510m$$
$$d = \sqrt{8670} \; km,$$
$$d = 93.11 \; km.$$

8.2.4 Constraints on Assigning Analog Channels

When allocating the four transmission frequencies for analog terrestrial television services from any mainstation site, it was desirable that where possible the consumer should only have to install a single aerial to receive all of the services. However domestic UHF receive aerials tend to have a limited frequency range of operation where antenna gain can be expected. In this case the receive antenna gain maximizes the electrical field that is coupled into the antenna. This then increases the electrical signal (terminal voltage) that is produced at the output of the lead connected to a television receiver, and within signal level bounds, it will produce a better quality signal.

As a result of the limited frequency response of receive aerials, it was necessary to impose constraints on the frequency separation of the broadcast channels used to transmit television signals from any transmitter mainstation to these UHF receive aerials. All the channels used must fit into a receive antenna group, that is the operating frequency range of the antenna. There are other constraints that were considered when allocating suitable channels for broadcasting analog (system PAL I) television to consumers on terrestrial networks. The following section describes some of the more important constraints, which are based on an 8 MHz wide broadcast channel.

1. A standard (PAL I) television receiver will have a local oscillator (LO) which is set at a higher frequency than the desired channel. This frequency is the sum of the Intermediate frequency (IF) and the receive channel, and tends to be approximately 40 MHz above the receive channel (The IF signal is approximately 39.5 MHz.) This means that the L.O. could interfere with other television receivers on the receive channel + 40 MHz (based on an 8 MHz channel), which are in close proximity. This prohibits usage of the receive channel + 5 channels.

$$LO = Rx\ channel + IF$$

$$IF = 40\ MHz\ (approximately).$$

2. An image frequency will be generated by unwanted mixing of the LO and the IF signals at the receive channel + 9 channels, prohibiting the usage of this channel.

$$1^{st}\ Image\ channel = LO + IF$$
$$= Rx\ channel + 9\ channels$$

3. A 2nd image channel will similarly be generated at the receive channel-9 channels, due to mixing of the signals. Thus any receive channel selected prohibits the usage of that channel - 9 channels.

$$2^{nd}\ image\ channel = -(LO + IF)$$
$$= Rx\ channel - 9\ channels$$

4. To allow for practicable realization of high power channel combiners and filters at UHF channels, it is helpful to have some degree of channel separation between the four transmitter frequencies at the same mainstation. This helps to increase the isolation between transmitters and reduces interference from one service to another. This constraint in practice means that adjacent channels are not used.

As an example of the application of the above abstract analog channel constraints, a typical high power UHF transmitter station example is shown below in Figure 8.1. The station is located at Cairn Hill, Longford, in Ireland and has at present, four analog UHF 20 kW transmitters. The services are known as RTE1, Net2, TV3, and TG4. The channels used are shown together with the channels that are not available for usage once the first channel (channel 40 in this case) and subsequent channels were selected. It is a typical UHF channel grouping and this type of channel arrangement is to be found at many similar terrestrial sites.

8.2.5 DTV Channel Possibilities

As it is desirable that the consumer should be able to receive digital television signals with the same receive antenna installation as is currently used to receive analog television signals, suitable channels must be picked from the same channel group as the existing analog services for any particular mainstation site. This means that the broadcaster or network provider must use some of the channels that were not available for analog television usage. The most suitable of these so-called taboo channels for usage are the upper and lower adjacent channels to the existing analog television services. This is because the original constraints imposed upon

Channel No:	40	41	42	43	44	45	46	47	48	49	50
Cairn Hill Tx:	RTE1	--	--	Net2	--	--	TV3	--	--	--	TG4
adj. Ch:		--	40	43	--	43	46	--	46	--	50	--
L.O.(Ch. + 39.5MHz)	--	--	--	--	--	40	--	--	43	--	--
Image Ch (LO+IF)	--	--	--	--	--	--	--	--	--	40	--
Image Ch -(LO+IF)	--	50	--	--	--	--	--	--	--	--	--

Figure 8.1 Analog UHF channel assignment example.

the usage of adjacent channels were placed to aid the broadcaster in achieving greater transmitter isolation. It is not a problem for receivers to select adjacent channels provided the relative power level of the adjacent analog transmitter does not overload the receiver front-end (RF tuner). As we will discover later in this chapter most broadcasters can use the adjacent channels for DTV broadcasting provided suitable filters and combiners are used. In Section 10.9.2 and 10.9.3, we will see how broadcasters can use frequency off-sets and other techniques to provide more transmitter isolation when using adjacent channels for DTV at existing analog transmitter sites.

8.3 Digital Television Transmitters

Digital terrestrial television transmitters tend to be of two main types, solid state and inductive output tube (IOT). However tetrode transmitters could also be used for high power terrestrial broadcasting. Broadcasters will generally transmit terrestrial DTV signals at lower power levels than the existing analog television peak powers. This is because the required signal strength for reception of DTV is lower than that required for reception of analog television (see Chapter 11). Hence it is expected that the output powers required by most terrestrial DTV transmitters will be achieved using solid state technology. The general architecture of a DTV transmitter is given in Figure 8.2. It consists of a modulator, followed by some form of linearity corrector to counteract any distortions to the DTV signal during amplification. Next the signal is mixed up to the desired RF channel frequency using an up converter, and then the signal is finally amplified to the desired transmitter power level. We will examine these functions in more detail later in this chapter.

8.3.1 Typical Transmitter Output Powers

In the UHF band current solid state amplifiers that are used for DTV are capable of producing maximum transmitter output powers of typically

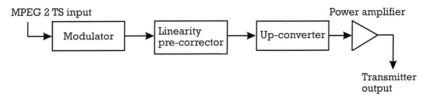

Figure 8.2 DTV transmitter functional diagram.

5 kW (rms), however some manufacturers can currently provide solid state transmitters with up to 8 or 9 kW output power. Above this power level the recommended approach is to use IOT technology, where transmitter powers of up to approximately 12 kW are quoted. In the VHF band solid state amplifiers are capable of producing maximum transmitter output powers of typically 5 kW (rms), however some manufacturers can provide solid state transmitters up to 12 kW. These figures are based on recent estimates from some manufacturers, and the achievable output powers are expected to increase as the technology matures.

8.3.2 Solid State Transmitters

The general architecture of a solid state transmitter is based upon the use of many parallel amplifiers, each of which is fed with part of the RF DTV signal and amplifies it. Each amplifier alone adds a small amount to the overall output power of the transmitter, typically about 300W. If any amplifier should fail, then the overall transmitter power will be reduced somewhat, but the transmitter will stay in operation (termed *on-air*). This is a significant benefit to the broadcaster in terms of system reliability, as redundancy can be engineered into the system design of the transmitter in terms of the number of amplifier modules used. Also the reliability of solid state amplifiers has greatly increased in recent years, with manufacturers quoting very long expected lifetimes in terms of the mean time between failure (MTBF) for amplifier modules.

Solid state amplifier modules are powered typically from low voltage, high current DC power supplies. The main supply voltage is generally a three-phase supply at standard voltage levels. This means that there is now often no need for dangerous high voltage supplies to the transmitter, as was the practice in the past with klystron-based devices.

8.3.2.1 Solid State System Configurations

Very many transmitter configurations are possible to select from, and manufacturers tend to offer variations of proven solutions to customers upon request. The more common configurations include the most basic type known as, single operation, and others offering varying degrees of redundancy including, dual drive operation, dual transmitters in parallel operation, dual transmitters in passive reserve operation, and dual transmitters in either passive reserve or active reserve operation.

Single operation means that the transmitter consists of the most basic configuration with minimal redundancy. The transmitter upconverter and pre-amplification processing is performed in the exciter unit, which

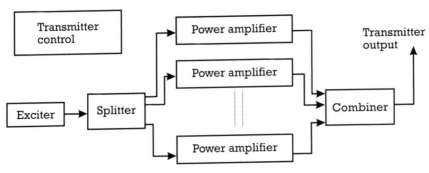

Figure 8.3 Solid state transmitter functional diagram, single operation.

in this case is single-ended (not duplicated) (see Figure 8.3). After splitting the RF DTV signal equally among the transmitter power amplifier modules, the DTV signal is amplified and combined again before outputting the final signal. There is some form of transmitter control and monitoring incorporated into the system.

Passive reserve operation is a variant of single operation, where there are now two separate exciter units (main and standby) which can be switched in or out of the transmitter amplification chain if the main exciter fails, by using an exciter change over switch, as shown in Figure 8.4. The changeover unit has a default condition to switch the main exciter into the amplifier chain if it should fail. This configuration has the advantage of removing the risk of exciter failure disrupting the transmitter operation, as can happen with the single-ended exciter of the single operation type.

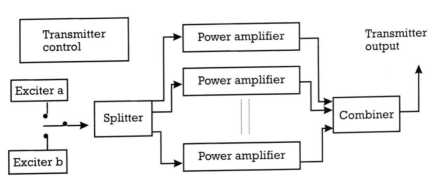

Figure 8.4 Solid state transmitter functional diagram, dual drive operation.

A third possible configuration is where there are two separate transmitters in parallel operation, as shown in Figure 8.5. This involves basically a completely redundant transmitter, which operates into a test load (also known as a *dummy load*) in normal operating conditions, simulating the antenna system. If the main transmitter fails then it is switched out of the main antenna system and into the test load while the redundant transmitter is switched from the test load into the antenna system. There are two exciters in this configuration, either of which can be switched into the two transmitter power amplifier sections, should any exciter unit fail.

Dual transmitters in passive reserve operation means simply that there are two completely independent transmitters fed with the same DTV signal, as shown in Figure 8.6. One transmitter operates into the antenna system while the other operates into the test load. In the event of a failure of any part of the main transmitter or exciter, the standby unit is switched into the antenna system while the main transmitter is switched into the test load. This system does not employ an exciter changeover unit and hence the two transmitters are more decoupled than in the case of dual transmitters in parallel operation shown in Figure 8.5.

Dual transmitters in passive and/or active reserve operation is the most sophisticated of the system configurations described, and is shown in Figure 8.7. It is used to allow for a special antenna fault condition

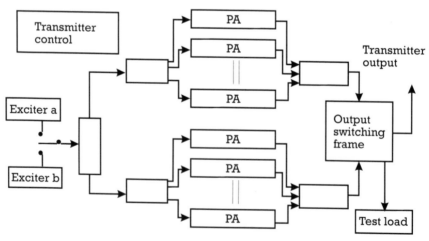

Figure 8.5 Dual transmitters in parallel operation.

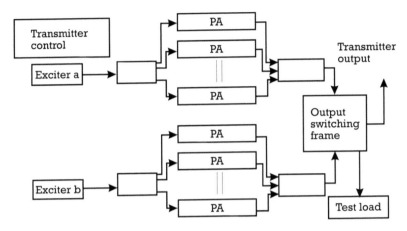

Figure 8.6   Dual transmitters in passive reserve operation.

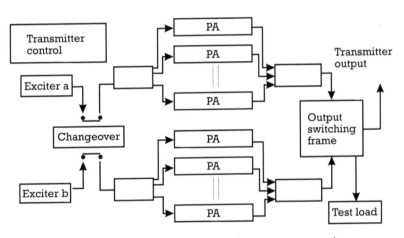

Figure 8.7   Dual transmitters in passive or active reserve operation.

known as half antenna operation, when for any reason only half the antenna system is available to transmit power. When this happens the ERP (emitted radiated power) of the mainstation will be reduced by 3 dB, and the launched power into the air will be reduced by half. This will cause some viewers close to the edge of the service area to lose reception of the DTV signal completely, a phenomenon known as the cliff effect. As this is not a desirable condition, the output power of the two transmitters can be added together (by a combiner contained within the switching frame) to increase the ERP by 3 dB (output power is then doubled). This compensates for the loss in efficiency of the antenna during this half antenna (or half stack) operation. The normal operating mode is the same as that described in Figure 8.6. Note that during half antenna operation when both transmitter amplifier chains are used in parallel, one exciter only is used to feed all of the transmitter power amplifier modules. This single exciter is selected to ensure that both transmitters produce identical high power signals with the same frequency and phasing, which could not be guaranteed if dual exciters (without phase combining) were used to drive both amplifier chains separately.

8.3.2.2 DTV Power Amplifier Classes

Power amplifiers are used in DTV terrestrial transmitters to deliver high power levels to loads, in this case the load is an antenna system (with a typical characteristic impedance of 50 Ω). The frequency response required for UHF systems is up to 860 MHz. The power amplifier itself must be capable of dissipating large amounts of power when operated at high current levels. There are two main classes of power amplifiers currently popular for use in DTV transmitters, class A, and class AB. Both bipolar junction transistors (BJT) and field effect transistors (FET) types can be used. MOSFET (metal-oxide semiconductor FET) transistors are becoming the most popular choice for power amplification because of the high gain that the devices exhibit and the superior reliability that can be achieved with these devices. Both BJT and MOSFET devices can be operated in class A or class AB mode.

An amplifier is said to be class A if its output remains undistorted during a complete cycle of a sine wave input signal. To do this the amplifier must be biased so that the output remains in the active region of the transistor characteristic curve irrespective of the value of the input signal. Class A amplifiers are not very efficient and the best efficiency achievable is typically very low, at about 1/4 of the total power consumed by the amplifier [2]. However, they are very linear and used to produce low RF output powers (up to 300W typically).

Class AB amplifiers are very commonly used at high power levels, and use two transistors in a push-pull configuration to drive a (possibly transformer coupled) load. The transistors are biased slightly to reduce any distortion that might be formed in the output signal whenever the input signal is zero valued. However the class AB amplifier is not very power efficient as both transistors are always in operation and as a result dissipating power.

8.4 DTV Transmitter Architecture

As DTV transmitters evolve, four main sections of the transmitter are emerging, as shown earlier in Figure 8.2. These include: the modulator, the linearity precorrector, the RF upconvertor, and the amplifier sections.

8.4.1 Modulator

All DTV transmitters will accept an MPEG-2 TS input signal as described in Chapters 6 and 7. The modulator will generate the channel-encoded signal, a COFDM, QAM, QPSK, or VSB signal. This will then typically be output at an intermediate frequency (IF) in a similar manner as analog transmitters currently operate. Processing at the IF frequency simplifies matters for the transmitter. The IF signal is then passed to the next part of the transmitter. However it should be noted that state of the art modulators can perform pre-correction before outputting the IF signal, and others process the DTV signal as I and Q signals for adaptive precorrection, without using an IF signal at all. These techniques are described in some detail in the next sections. Consult the references for more information about this subject.

8.4.2 Linearity Precorrector

This device which is becoming more and more incorporated within the modulator, has the function of pre-distorting the DTV signal at IF prior to upconversion and amplification. By having some knowledge of the distortion introduced during amplification the pre-corrector can correct for it, either in advance or as a dynamic process. The state of the art currently is in the usage of adaptive pre-correction methods on the I and Q components of the DTV signal and also using digital linearizing methods on the baseband digital signal prior to the RF amplifiers [3]. These digital adaptive baseband pre-distortion techniques can be applied to any DTV signal, independent of the modulation type employed. These techniques allow the transmitter to generate higher output power levels with acceptable signal distortion.

8.4.3 RF Upconvertor

This piece of equipment is very similar to analog transmitter upconversion circuits. The DTV signal is mixed with a local oscillator (LO) to generate the desired RF channel and then filtered to remove any unwanted sidebands generated in the mixer. The LO phase noise must be kept within a tight specification, as DTV signals are sensitive to phase noise degradation. If precision offset is to be used to offset the RF signal from the designated channel center frequency then the upconvertor circuit must be capable of frequency locking to a frequency reference signal. For single frequency network (SFN) operation it will also be necessary to lock the MPEG-2 TS input signal to a timing reference pulse such as a PPS (pulse per second) signal (see Chapter 10 for more details).

8.4.4 Power Amplifier and Derrated Operation

This is a very important part of the DTV transmitter and it was discussed earlier in Section 8.3.2.2. However from the overall transmitter perspective the most important parameter is the amount by which the transmitter amplifier must be derrated from normal analog power operation. This is sometimes termed as the output back-off (OBO) and is a measure of how much the transmitter must be reduced from the maximum output power it could achieve (using an analog peak sync signal). However, this commonly used definition is not strictly correct because a peak sync power definition is not a measure of the average power an analog transmitter produces. Nevertheless manufacturers often quote it.

8.4.4.1 Output Back-off definition

The more formal and more correct definition for OBO is that:

$$OBO = 10 Log \frac{Psat}{Po} \qquad (8.2)$$

where *Psat* is the maximum average output power the amplifier can produce (saturation power) with an unmodulated carrier wave (CW) at the input, and *Po* is the average power it actually delivers at the operating point [4].

8.4.4.2 Crest Factor

Another parameter commonly used when discussing the peak and average power levels of a DVB-T digital signal is the crest factor, which relates the ratio of the peak power of a multiple carrier signal to the average power of the signal. It is also referred to as the peak to average power ratio (PAP) [5].

$$\text{Crest factor} = 10 \log N \text{ (dB).} \qquad (8.3)$$

where N = the number of carriers.

For DVB-T systems where there can be approximately 6,800 carriers used for the 8k mode of operation, there could be statistically a power crest factor of 38 dB. Fortunately the likelihood of constructive addition of the many thousands of carriers is very low due to the necessity for correct phase addition. It is reported that a DVB-T signal can be characterized by a 12 dB crest factor correctly for 99.99% of the time [6]. However it is important to allow for the crest factor of a multicarrier signal when specifying components which might carry more than one DTV signal, such as a combiner, or antenna system.

8.4.5 DTV Transmitter Distortions and Clipping

Amplification of any digital signal will cause some generation of nonlinear distortions because of the broadband nature of DTV. COFDM signals, as was seen in Chapter 7, contain literally thousands of carriers and the time domain signal has peak excursions that are orders of magnitude greater than the average signal level. These peaks can drive an amplifier into saturation and generate nonlinear distortions, unless the signal peaks are clipped. Clipping also introduces some degree of distortion, but it can be controlled and set to a particular level. By increasing the transmitter OBO the likelihood of driving the amplifier into a region of nonlinearity is reduced, and the level of clipping needed can also be reduced. However, increasing OBO reduces the transmitter efficiency and results in large transmitter sizes (footprint) for relatively small output power levels. The COFDM modulation scheme, which produces a time-varying signal envelope with a relatively high peak-to-average ratio, [7] is very sensitive to high power amplifier (HPA) nonlinearities, and attention should be taken to the reduction of these peaks.

COFDM intermodulation products (IPs) are generated in the power amplifier stages of DTV transmitters and the 3rd order intermodulation products predominate. The IPs roll-off to negligible levels at approximately ± 7.6 MHz either side of the RF channel spectrum edges, but this indicates why DTV channel filters are needed after the transmitters. The IPs are also present within the DTV spectrum and are typically quoted as dB relative to the DTV signal spectrum envelope.

8.5 Filters, Combiners, and Antenna Systems

As was mentioned in the previous sections, high power television transmitters produce unwanted emissions due to the nonlinearity of the high power amplification stages of digital television transmitters. These spurious emissions are present both in-channel and out-of-channel and can cause interference to adjacent channels. As was mentioned in Section 8.2.4, adjacent channels are being used extensively for hybrid digital/analog television networks to allow usage of common receive antenna systems for consumers. So in order to protect adjacent channels from interference, channel filters with sharp cut-off responses are needed.

8.5.1 DTV Filters

DTV filters need to have tight responses to ensure that there is a minimum of RF emissions into adjacent channels. As a result they tend to employ many cavities to ensure that the filter has a steep slope on either side of the channel.

For DVB-T signals that are intended for usage in 8 MHz wide channels, the signal bandwidth is 7.6 MHz wide. Typical insertion loss values are 0.2 dB in-channel and >45 dB ± 8 MHz from the channel center. Typical in-channel return loss figures are >23 dB across the channel.

8.5.2 DTV Combiners

For combiners, the main issues are the selectivity required for adjacent channel combining, the potential voltages generated within the combiner (see Section 8.5.4), and the group delay impact on any analog services due to the steep response of the cavity filters used for DTV.

Adjacent channel combiners are being deployed as a result of the limited spectrum availability for DTV services. These combiners must be able to reject and select two adjacent RF channels to provide isolation between the two transmitters. This is difficult to achieve, and requires using 6 and sometimes 8 cavity-tuned filters.

Group delay penalty is due to the sharp cut-off of the DTV filter response, which means that there is a large group delay incurred. This has little effect on a DTV signal such as the DVB-T signal, which can handle very large transmission delays, but it means that extra group delay correction may be needed for analog transmitters that are combined with DTV signals.

8.5.3 DTV Antenna Systems

DTV antenna systems are the same as those that are currently used for analog television broadcasting. The only requirements are that the antenna system be capable of handling the average (thermal power) of the DTV signal, and that the peak voltages of the combined signals do not exceed the voltage breakdown levels for antenna distribution cables and connectors.

8.5.4 Voltage and Power Ratings

Most systems that involve only a single DTV transmitter will be able to use similar (if not lower) rated coaxial feeder, and connectors as is presently used for analog systems. It is expected that DTV transmitter amplifiers will tend to saturate at the OBO level and limit the value of the peak powers and voltages generated to that level. However where more than one DTV signal is carried in a feeder or cable there can be quite large voltages generated whenever all the DTV signals add together in phase.

Example 8.2

In a six channel DTV system, where each transmitter is set to 1 kW (average power), what peak voltages could be expected? The OBO is assumed to be 6 dB for each transmitter.

> *Power = (Voltage).(Current),*
> *Voltage = (Current). (Impedance),Ohms law*
> *∴ Power = (Voltage)^2/ (Impedance),*
> *Here the feeder impedance = 50Ω.*

> *For each transmitter, the expected average voltage can be calculated,*
> *1,000 = (V)^2/(50)*
> *∴Voltage (average) = 223.6V, for each transmitter.*
> *So the total average voltage for six transmitters = 1,341V.*

> *Now if the OBO for each transmitter is 6 dB,*
> *then the peak power of each transmitter = 4 kW.*

> *For each transmitter, the expected peak voltage can be calculated*
> *4,000 = (V^2)/(50)*
> *∴Peak voltage = 447V*

> *Total peak voltage for 6 DTV signals is (447) · (6) = 2,682V!*
> *This potential peak voltage could be doubled during half antenna working, when each transmitter is operated at an extra 3 dB power level.*

As was seen earlier when discussing the crest factor of a DVB-T signal, the random nature of the signal means that it is necessary to use statistical analysis to predict the peak voltages and power excursions of the signal, that may occur if the time scale is long enough. That is, if the equipment is left in service for a long period of time. However if the above example is used as a basis for designing power ratings for equipment, then considerable over-engineering may be expected, with very large connectors, combiners, and costs! It is accepted as incorrect to assume that the duration of any peak voltages generated will be so short as not to cause gas breakdown in DTV transmission systems.

One alternative approach is to calculate the peak voltage that is likely to happen at a more regular period of time, possibly every 10 years, and design systems using factors of safety based on this occurrence frequency. The system designer must make choices about the likelihood of a peak voltage or power level being produced within the lifetime of the equipment in order to minimize the ratings and cost of equipment.

8.6   DTV Transmitter Size and Cooling Arrangements

In recent years many analog solid state transmitters have been using air-cooling systems to remove the excess heat from the power amplifier modules. Very high amplifier reliability has been achieved using well-engineered air-conditioning plants. By reducing the junction temperature of an amplifier transistor by even a few degrees it is possible to extend the lifetime of the transistor considerably. It is therefore very important that a solid state transmitter is adequately ventilated to prolong the lifetime of the amplifier modules and reduce the likelihood of transmitter failure. Air-cooled transmitters are large because of the large surface area needed to dissipate the excess heat from the amplifier modules and the large ducts needed to bring/remove air from the transmitter. In Table 8.1 below, some typical manufacturer specifications for air handling requirements for transmitters are given.

8.6.1   Air Cooling Systems—Disadvantages for DTV

Air-cooled transmitters generate a lot of unwanted acoustic noise due to the fans that are used to blow air over the amplifier heat sinks. Also the ducting used to bring cool air to the transmitter and to remove hot exhaust air from the transmitter hall can be quite large and cumbersome.

Table 8.1
Typical Analog Transmitter Air-handling Requirements

Analog transmitter power (peak sync)	Volume of air required (approximate values)
10 kW	150 m³/min
20 kW	200 m³/min

Solid state amplifiers require large quantities of cool air to ensure that the amplifiers remain in the normal operating region, if the requirements are not met the temperature of the transmitter amplifier modules will rise quickly and amplifier failure is inevitable. It can be quite difficult to provide the required quantities of dry and cool air for the transmitter, at certain times of the year and at high altitude, where many terrestrial transmitter sites are located.

8.6.2 Liquid Cooling Systems

Water or liquid cooling is becoming attractive in the DTV transmission era, because transmitter halls are generally full of analog transmitters and there is little room to accommodate large air-cooled DTV transmitters. Liquid cooling technology has recently been developed to achieve high reliability with solid state amplifiers. Liquid cooling uses pumps to bring a coolant liquid (which is usually water) to the amplifier heatsinks and then the liquid is warmed by the heatsink removing the thermal energy from the amplifier. The warmed liquid is removed to an external cooling unit to remove the thermal energy. It is a virtually silent process unlike air-cooling, which is very noisy, and is attractive therefore from a health and safety viewpoint. Liquid cooling can be used to heat parts of the building in winter that normally require space heating, by using a heat condenser. Because there is no need for large fans, ducts, and the air handling rooms to filter and prepare the air for the transmitters, considerable space saving can be achieved. Also liquid cooled transmitters tend to be smaller than equivalent power air-cooled transmitters. As a result of these considerations, liquid cooling is at present the focus of attention for many broadcasters and manufacturers.

8.6.3 Geothermal Cooling Systems

Some network providers are currently examining geothermal cooling in conjunction with liquid cooling as this arrangement does not require an active cooling unit to remove waste heat from the liquid. Geothermal systems operate independently from the prevailing climatic conditions, as the ground temperature is stable throughout the year. Geothermal technology is advanced in the areas of building climate control, especially in the United States; however, it has only recently begun to be deployed in some broadcast transmission systems.

References

[1]   Hutson, G., P. Shepherd, and J. Brice, *Colour Television, System Principles, Engineering Practice and Applied Technology*, New York: McGraw-Hill, 2nd ed., 1990.

[2]   Bogart, T., *Electronic Devices and Circuits*, Merrill, 1986.

[3]   Andreoli, S., and H. McClure, "Digital Linearizer for RF Amplifiers," *IEEE Trans. On Broadcasting*, Vol. 43, No. 1, March, 1997.

[4]   Le, M., and L. Thibault, "Performance Evaluation of COFDM for Digital Audio Broadcasting, Part : Effects of HPA Nonlinearities," IEEE Trans. On Broadcasting, Vol. 44, No. 2, June, 1998.

[5]   Wang, X., et al., "Reduction of Peak to Average Power Ratio of OFDM System Using a Companding Technique," IEEE Trans. On Broadcasting, Vol. 45, No. 3, September, 1999.

[6]   Dittmer, T. W., "Digitally Modulated RF Systems," Proc. IBC 1998, Amsterdam, pp. 215–227.

[7]   Schilpp, M., et al., "Influence of Oscillator Noise and Clipping on OFDM for Terrestrial Broadcasting of Digital HDTV," *Proc. IEEE Int. Commun. Conf. 1995*, Seattle, USA, pp. 1678–1682.

Contents

DTV Receivers and Measurements

9.1   Introduction

This chapter will deal with aspects of the reception of digital television including measurement procedures to quantify the quality of the transmitted and received signal. It will focus on terrestrial reception aspects, however many of the techniques are equally applicable to other transmission media. Common types of signal degradation will be illustrated. As the definition of a DTV receiver can include both set top boxes (STB) and integrated receiver decoders (IRD) this chapter will also discuss the functionality of these set top boxes and integrated receiver decoders. An introduction to the topic of software to control the working of the receiver, that is, middleware and application programming interfaces (API), will be given together with a discussion on electronic program guides (EPG).

A functional diagram of a standard DTV receiver is shown in Figure 9.1 for terrestrial systems. It consists of a radio frequency tuner or front end, followed by an analog to

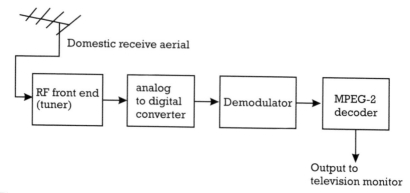

Figure 9.1   Terrestrial DTV receiver functional diagram.

digital converter, next a (COFDM) demodulator is shown, and finally, an MPEG-2 decoder unit. This generic system is similar to those used for other transmission media, except for the usage of different RF front ends at different broadcast frequencies, and also that different types of demodulators are used with the different transmission media modulation schemes, such as 64-QAM, QPSK, VSB, and COFDM.

It is expected that by the end of the 1st year of the new millennium there will be over 30 million DTV receivers in usage throughout the world, and that approximately 10 million of these will have some degree of interactive capability. By the year 2003 over 60 million receivers are expected to be deployed, and the majority of these will provide the consumer with interactivity. In Europe to date over 80% of all set top boxes deployed have an interactive capability while only 5% of the set top boxes in the United States are interactive. Clearly the industry is divided about what the differences between analog and digital broadcasting will be, in the new millennium. The European terrestrial network operators are beginning to voice the opinion that the key difference will be interactivity, not the quantity of channels offered nor the digital signal quality. Another important development for DTV receivers is the increased availability of mass storage at low cost, to allow the downloading of applications and data to the receiver. As was mentioned earlier in Section 5.6, this can be done at optimum times to maximize the available network bandwidth usage. For instance services such as movies could be downloaded to the receiver at off peak viewing times, when the network provider could make more resources available for this service. This has great potential in bandwidth limited environments such as terrestrial networks which can now compete with other transmission media in delivering movies and data to local receivers for consumption at a later and more convenient time.

9.2   Measurements

With digital television transmission systems, it is possible to perform many different types of measurements on either the digital bitstream prior to modulation, or on the modulated signal itself [1]. Some of these measurements are completely new to broadcasting, however many of the RF measurements are similar in methodology to those employed in analog broadcast systems. Among the more common measurements performed in DTV broadcast systems include, measuring the bit error rate (BER) of a received and decoded signal, measuring the carrier-to-noise ratio (C/N) of the modulated signal, and performing a constellation analysis of the in phase and quadrature phase components of the modulated signal in the time domain. Also of importance in broadcasting is the measurement of the true signal power, the level of phase noise of any local oscillators (LO) used in transmission and reception, and the linearity of any transmitters used to broadcast the DTV signal (see Chapter 8). Also of emerging importance is the measurement of the modulation error ratio (MER), which can give a figure of merit analysis of the COFDM ensemble. A brief summary of the more common measurements and definitions now follows.

9.2.1 RF and IF Signal Power

The power level of a digital television signal is measured usually at the RF stage or at the IF stage within a DTV transmitter. In any digital television system the signal power is defined as the mean power of the signal measured with a thermal power sensor. It is important to measure the power within the complete bandwidth of the signal at the frequency of interest. It is recommended to use a calibrated channel filter at the output of a transmitter to ensure that only the signal in-channel power is measured. If a spectrum analyzer is used to measure the channel power, it is common to integrate the power measurement over the signal bandwidth.

9.2.2   Noise Power

Noise power within the channel is a very important measurement, and is used in order to calculate the carrier-to-noise ratio of the system when designing transmission networks or installing equipment. It is necessary to measure the in-channel noise level, and this can be done for COFDM signals by removing some of the signal carriers to create an in-channel hole in the spectrum. It can also be performed by measuring the noise floor for the channel without the presence of the carrier ensemble. The equivalent noise bandwidth must then be integrated over the complete channel bandwidth from this measured value.

9.2.3 Carrier-to-Noise Ratio (C/N)

The signal power (carrier) is first measured as described above and then the noise level (mean power) is measured using the equivalent noise bandwidth of the OFDM signal. To obtain the C/N value the ratio of the two figures is expressed as the C/N in dB.

In practice if it is possible to switch off the transmitter and select manual gain control with the receiver, then the difference between the spectra as observed on a spectrum analyzer with the power on and power off is a simple measurement method. Otherwise the noise bandwidth method can be employed. It is anticipated that the final DVB-T measurement guidelines document will address this topic in more detail [2].

9.2.4 BER Versus C/N

This is a very useful measurement and among the most commonly used to indicate coverage and signal quality. It can be measured, at a receive site, to give an indication of receive signal quality or at a transmit site where it is used to evaluate the performance of a transmitter and is useful for setting the transmitter back-off level. A pseudo random binary sequence (PRBS) is usually input to the modulator, upconverted to the channel frequency, amplified and broadcast. For this reason a test transmitter should be able to generate a standard telecommunications $(2^{23}-1)$ PRBS signal as defined by ITU-T recommendation 0.151 [3]. There are various types of PRBS signals in use with DTV, but it is best to use telecommunications approved signals. The received signal BER and C/N are measured for a variety of transmitter output power levels. In normal operation it is not possible to replace the MPEG-2 TS with a PRBS signal as the transmitter is in operation with live traffic, hence another method to indicate the BER from the MPEG-2 TS is used in some professional DTV receivers. Measurement methods based on software BER calculations have recently been developed and have been shown to approximate quite closely the measured BER when using a standard telecommunications test BER signal. When measuring the BER of DTV signals using for example DVB-T receivers, it is possible to perform the measurement at different points within the receiver, and these options will be compared later in Section 9.4.

9.2.5 Transmitter Output Back-off

The transmitter output back-off is defined as the ratio of the peak sync power output of the transmitter (when broadcasting a pulsed power signal, such as an analog TV signal) to the transmitter mean output power when broadcasting a DTV signal. It is measured in decibels (dB). It is used as a rule of thumb to give an approximate indication of the relative size of the transmitter required to produce a DTV signal, compared to an analog TV signal. It also is used as an indicator of the transmitter performance. However as DTV systems become more widespread, this parameter should become less relevant when comparing transmitters.

9.2.6 Phase Noise

Phase noise due to any unstable local oscillator in the transmission chain can cause common phase error (CPE) which leads to a circular smearing of constellation points and also inter-carrier interference (ICI) which is noise like in appearance. While it is possible to precorrect for CPE, phase noise should be minimized as ICI cannot be corrected for at present. A common source of ICI is an interference signal at a precise frequency. This type of signal is often referred to as spurious sine, having most power concentrated in a single frequency (jammer), and it can cause a strong disturbance of adjacent OFDM carriers.

In specifying a complete transmission system a maximum tolerable phase noise characteristic is often set, for certain bandwidth offsets from the carrier signals used within local oscillator circuits. This phase noise mask is generally defined for the system and is described in terms of dBc/Hz at a certain frequency offset from local oscillator. The term dBc/Hz refers to the phase noise level in decibels relative to the carrier, at a frequency offset from the carrier.

9.2.7 Transmitter Linearity

DTV signals in the frequency domain have characteristics that often resemble bandpass filters, with steep shoulders at the edge of the signal. Transmitter linearity is characterized by measuring the shoulder attenuation of the RF signal spectrum, and is measured in dB. This involves measuring the maximum value of the signal spectrum on a spectrum analyzer and then subtracting a value measured 500 kHz away from the upper and lower edges of the signal spectrum. This gives a measure of the intermodulation products at the spectrum edge. Typical values of shoulder attenuation or distance are more than 40 dB below the main signal level. Higher performance can be achieved with pre-correction techniques within the transmitter exciter.

9.2.8 Equivalent Noise Degradation (END)

This is defined as a measure of the implementation loss (reduction in carrier-to-noise performance) caused to an ideal transmission system by the addition of any piece of real equipment. The normalized value (reference value) is that assumed for an ideal network. Typical values are in the order of 0.5 dB END value for DTV transmitters.

9.2.9 Modulation Error Ratio (MER)

The modulation error ratio is a method of calculation of a figure of merit for a vector-modulated signal (such as a COFDM signal). It is expressed in dB and allows a receiver to give a figure of merit for the COFDM carrier ensemble. It is becoming popular within professional DTV receivers and is used in a similar manner to the carrier-to-noise ratio.

9.2.10 Quality of Service

Generally with analog television, the quality of service (QoS) of a television program is determined by the length and quality of the transmission path between the transmitter and a domestic receiver. However with the move to digital broadcasting the quality of a program will now be more determined by the encoding process at the beginning of the transmission chain rather than the transmission path. With digital modulation schemes and digital error correction facilities it is possible to guarantee error free reception at a domestic receiver. The MPEG-2 encoding process, which is usually at the beginning of a transmission path, uses compression and as we have seen this may reduce the quality of the signal.

There is a requirement to perform real time measurements of MPEG-2 picture quality without access to a reference source. These measurements make use of subjective criteria to grade the quality of a digital video service. To make subjective quality ratings the ITU has specified at least two test methods to date. These include the double stimulus continuous quality scale (DSCQS) method and the single stimulus continuous quality evaluation (SSCQE) method. The DSCQS method is used for comparative assessments while the SSCQE method is based on a single observation of the sequence to be assessed. Both methods use a scale from 0 to 100 to cover the quality levels excellent, good, fair, poor, and bad. Both methods take into account the subjective nature of the human eye to degradations and picture movement.

9.3 Time Domain (IQ) Signal Analysis

Time domain signal analysis can be used in DTV systems and the most common involve constellation analysis of the I (in phase) and the Q (quadrature phase) components of the DTV signal. In phase and Quadrature phase (I/Q) analysis is commonly used in single carrier DTV systems (see Chapter 6) and it can equally be applied to the DVB-T COFDM multi-carrier system. This is because the COFDM system utilises typical digital modulations such as QPSK, 16-QAM, and 64-QAM to map data onto each carrier (see Section 7.7 for a description). To use this analysis it is necessary to have access to the I and Q components of the signal within the demodulator. Groups of carriers can be superimposed and treated collectively, or the analysis methods can be applied to all of the carriers taken together. Some of the more common measurements are outlined below. As a reference a typical error free 64-QAM constellation is shown in Figure 9.2 below. This constellation waveform is generated using an oscilloscope by plotting the I component as a function of the Q component (using the oscilloscope XY function). Note that an actual DVB-T signal constellation would also include constellation points generated from the presence of continuous pilot carriers (see Section 7.9). In Figures 9.2–9.5 the actual constellation point sizes have been exaggerated to highlight the effect of the error on a typical constellation waveform.

9.3.1 Phase or Quadrature Error

If the phases of the two carriers feeding the I and Q modulators are not orthogonal (90 phase difference) then the resulting phase error will generate a constellation where the lines formed by the I/Q pairs are no longer parallel to the decision thresholds, as illustrated in Figure 9.3 below. This reduces the noise immunity of the system.

9.3.2 Phase Jitter

A QAM signal with superimposed oscillator phase jitter will form a constellation with circular segments as shown in Figure 9.5 . Note that the central segments are the shortest, as a result these segments have the

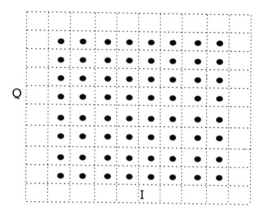

Figure 9.2   Typical error free 64-QAM constellation diagram.

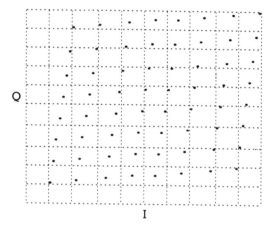

Figure 9.3   64-QAM signal with constant phase error.

least sampling uncertainty. However the outer segments are more prone to receiver error as they are the longest and as a result are subject to more sampling uncertainty in the receiver.

9.3.3 Interference

A QAM signal subjected to interference can generate a constellation where the points form complex shapes, such as circles (as shown in Figure 9.5), ellipses, and Lissajou figures.

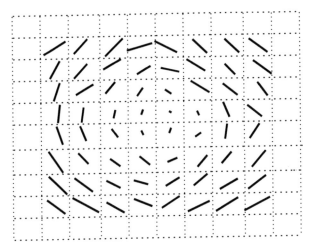

Figure 9.4   64-QAM with phase jitter.

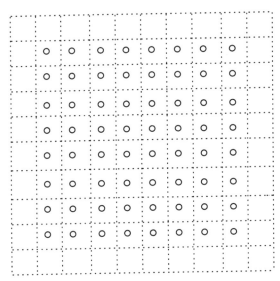

Figure 9.5   64-QAM constellation with an interference signal.

9.3.4 Noise

A QAM signal with added noise forms a constellation diagram where the pairs of I/Q values form symbol clouds. As the noise level increases the constellation points are enlarged and there is a greater probability of sampling uncertainty within the receiver. Hence the likelihood that the receiver will incorrectly decode a constellation point increases.

9.4 Measuring Bit Error Rates (BER)

Bit error rate is the most important parameter used in assessing the quality of a digital transmission channel, as it is a direct measure of the number of bits that have been received in error by the system receiver. Bit error rate can be defined as the ratio between erroneous bits and the total number of transmitted bits. By using a known sequence of bits as a reference, it is possible to compare the transmitted and received sequence of bits, including any errored bits, to a known sequence of bits without errors, and from the two calculate the BER. This can be done using a known test sequence of bits such as a PRBS signal (see Section 9.2.4).

9.4.1 BER Measurement Points for DVB-T Receivers

As mentioned earlier there are a number of points within a DTV demodulator (receiver) at which it is possible to measure the bit error rate. Specifically for a terrestrial DVB-T receiver it is possible to measure the bit error rate at three distinct points (see Figure 9.6).

1. Measurement point 1 is measured before the Viterbi decoder.

2. Measurement point 2 is measured after the Viterbi decoder and before the Reed Solomon decoder.

3. Measurement point 3 is measured after the Reed Solomon decoder.

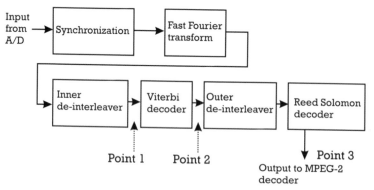

Figure 9.6 Terrestrial demodulator functional diagram.

BER is usually plotted as a function of the receiver measured carrier-to-noise ratio (C/N), as this is the most common measurement made to quantify the quality of a transmission path.

9.4.1.1 Measurement Point 1

As mentioned above point 1 corresponds to measuring the BER before the Viterbi decoder. At this point it is possible to get an indication of the in-service uncorrected bit error rate. It is useful for giving an indication of the quality of the transmitter, the transmission channel, and the receiver tuner and analog to digital converter. It corresponds to a raw BER, as it does not allow for any functions within the receiver that use digital error correction techniques to remove errored bits. It results typically in very high BER measurements that vary little in absolute magnitude as the C/N ratio varies. Typical values of BER measured at point 1 would be in the range of 1.0E-1 to 1.0E-2, which is from 1 error for every 10 bits transmitted to 1 error for every 100 bits transmitted. These correspond to very high bit error rates, which would cause unacceptable picture quality for use in a DTV receiver, and this illustrates why powerful error correction techniques are used within all receivers. However it results in a very shallow curve when BER is plotted against C/N ratio, which can be contrasted to the steep curves generated when measurements are made at measurement point 2. As an illustration of the variation of BER measured against C/N (see Figure 9.7) where typical curves are shown corresponding to measurement points 1 and 2. The curves correspond to those measured for a 64-QAM, code rate = 2/3, COFDM signal. Note that a Gaussian

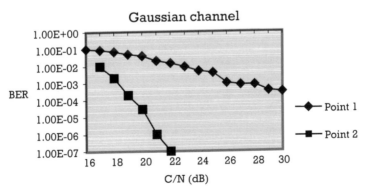

Figure 9.7 Bit error rate versus carrier-to-noise ratio as measured at points 1 and 2 in a DTV receiver for a 64-QAM, rate 2/3, COFDM signal.

channel is defined as a channel where the noise impairment present is assumed to be white-noise like, that is randomly distributed throughout the channel.

9.4.1.2 Measurement Point 2

Measuring the BER at point 2 is the most commonly used reference measurement point for DTV terrestrial systems. At point 2, which is at the output of the Viterbi decoder and just prior to the Reed Solomon decoder, a target error rate of 1 errored bit per 5000 transmitted bits is defined as the quasi-error-free reception point (QEF). QEF corresponds to a BER = 2.0E-4 measured at point 2, and this results in an output BER of less than one error per hour, where output in this case corresponds to measurement point 3, that is after the Reed Solomon decoder.

Example 9.1

A DVB-T signal is transmitted using 64-QAM, code rate 7/8, and guard interval = 1/32, which corresponds to a useful bit rate of approximately 31.67 Mbps. What will the measured BER be at the output of the Reed Solomon decoder (point 3), if the BER measured at point 2 is found to correspond to the QEF point?

Net usable bit rate = 31.67 Mbps, and this corresponds to a bit rate of 114,012 Mb/hr.

114,012 Mb/hr = 31.67 Mbps × 60 (sec) × 60 (min).

QEF = 1 error/hr, measured at point 3 in a receiver, or a BER = 2.0E-4, measured at point 2 in a receiver.

In this example QEF = 1 error per 114,012 Mb transmitted.

$$\therefore \text{The BER (measured at point 3)} = \frac{1}{14,012 \times 10^6}$$

$$\therefore BER \approx 8.7\,E-12.$$

It can be seen from Example 9.1 above that the QEF bit error rate of 2.0E-4 measured at point 2 corresponds to an approximate bit error rate of only 8.7E-12 when measured at point 3, after the Reed Solomon decoder. In practice it would take a considerable amount of time to measure such a low bit error rate at point 3. This long measurement time is not practical, and this is one of the reasons why point 2 (instead of point 3) is used as a measurement point for transmission system quality.

However measuring the BER alone could mask underlying transmission problems if the carrier-to-noise ratio is not also measured. This is because digital systems employing Reed Solomon error correction can be close to total system failure and yet display very little residual errors. This phenomenon is due to the very steep nature of the characteristic BER versus C/N ratio curve for systems measured at point 2. It is often termed the cliff-effect as the BER is seen to fall from very high levels to extremely low levels within a few dB variation of C/N ratio. The huge variation of BER for a relatively small change in carrier-to-noise ratio means that a system, which is not showing significant degradations (low BER), could be on the verge of failure (on the cliff edge). If then, for any reason the carrier-to-noise ratio were degraded by a small amount the overall transmission channel could fail.

9.4.1.3 Measurement Point 3

Measurements made at point 3 are at the output of the demodulator and are the least useful for technical evaluation of the transmission system and channel. This is because these measurements are made after using all of the powerful error correction techniques that DTV can employ to remove errors. There is a risk that the error correction techniques could mask any channel impairments or transmission system degradations. Point 3 should not be used for transmission quality evaluation as it will inevitably lead to problems when link budgets become reduced due to climatic effects or transmission system degradations [4].

9.5 Set Top Boxes (STB) and Integrated Receiver Decoders (IRD)

A set top box (STB) or set top unit (STU) is a device which can translate digital television signals to a format that can be interpreted and displayed by existing analog television receivers. STBs are sometimes referred to as digital set top boxes (DSTB). It includes all of the hardware, middleware, and access entitlement software to allow the consumer to decode all the available digital video, audio, and data. It allows access to all of the digital based content and services using existing analog television receivers as the display unit. An integrated receiver decoder (IRD) contains also a display unit. Most IRDs available to date can only demodulate the terrestrial COFDM modulation, as traditionally receivers were manufactured to decode off-air or terrestrial signals. Also they are very new additions to the consumer electronics market and at present most can only decode unencrypted DTV signals. However they should become more popular as the DTV market consolidates.

9.5.1 Set Top Box Functionality

Set top boxes are widely available for DTV and in the case of the COFDM based DVB-T system, both the 2k and the 8k systems are now supported. A typical STB will connect to an existing analog television receiver using standard interconnections such as a SCART connector and/or an RF loop through connection, and in a similar manner connect to a video cassette recorder. (The predominant interconnection standard for domestic equipment is the SCART connector, which derives its name from the French committee that has promoted its use.) Many STBs use telephone modems as the return channel, however other terrestrial based systems will be deployed in the future for the return channel. Stereo audio in either digital or analog format can be output to, for instance, a HiFi system. The internal MPEG-2 decoder typically supports Main Profile@Main Level with data rates per service of up to 15 Mbps (see Chapter 4). Current STB units generally support widescreen video format. The hardware can be manufactured to support various application programming interfaces, and conditional access systems. At the moment the network operator financially supports most STB deployment and this is leading to the emergence of proprietary STB systems for different transmission media. In Europe the DVB common interface (CI) is not mandatory for STB. This interface allows for the interconnection of DVB equipment and subassemblies.

9.5.2 Integrated Receiver Decoder (IRD) Functionality

At present, a few manufacturers have begun offering complete DTV receivers, a configuration that is called an integrated receiver decoder. These devices allow for the reception and decoding of DTV without the need for a STB. Some offer flat screen displays and all can decode the unencrypted signals at the moment. They incorporate primarily the demodulation of the terrestrial based transmission standards.

9.6 Middleware and Application Programming Interfaces (API)

There is no doubt that one of the most important topics for broadcasters and service providers at the present time is the topic of set top box middleware and application programming interface (API). In order to provide data services the set top box requires a layer of software that resides between the STB hardware and the application which is to be consumed. This software is called "middleware" and is often referred to as an API when used in connection with DTV. The term API is commonly used in software engineering but in DTV engineering some middleware solutions incorporate the API, but some do not. Some commentators refer to the API as the equivalent to "Windows" for the STB. It can be a helpful analogy for those new to this terminology.

It should be noted that STBs having different middleware might not be able to access the same data or interactive services. Each service must be designed or authored for each particular middleware system that it is intended to run on. This is a major problem for traditional analog broadcasting organizations entering the DTV market [5]. Clearly a fragmented STB market with different STB manufacturers using different middleware will mean that the same material will need to be authored many times to run on different STBs. This is recognized as a potential disaster for DTV and as a result DVB wish to adopt a standard for middleware known as the DVB multimedia home platform (DVB-MHP). This API will meet the need for the next generation of STB, including interactive applications, and Internet access from the STB. Other middleware solutions are proposed in the meantime until the DVB-MHP is developed and standardized. It is beyond the scope of this book to explore the issues relating to middleware and APIs.

9.6.1 DVB and Java

The DVB group has decided to use the Java programming language as the core specification for the software of the DVB-MHP (multimedia home platform). Java is a powerful programming language that should allow the implementation of new applications in a platform independent manner. Providing a so-called virtual machine environment, that is, a software entity that processes applications in the same way on any microprocessor in which it is implemented, does this. This will allow many different hardware realizations of the same applications. With Java, the virtual machine is commonly called a Java virtual machine (JVM).

9.6.2 The MHEG API

MHEG is a sister organization of MPEG within ISO/IEC JTC1 (see Section 3.2). The multimedia and hypermedia information expert coding group (MHEG) develops standards primarily for the transmission and representation of data and applications to allow for interactive services on set top boxes. The MHEG-5 standard has been chosen in the United Kingdom implementation of DTV and is now in use. It is an open and nonproprietary standard for the set top box API.

9.6.3 The EuroMHEG System

The EuroMHEG is a development of the MHEG-5 API to include extensions to the standard. The purpose of these extensions is to provide greater functionality to the consumer, within an open standard API. These functions include provision for a return path in various ways, and a financial toolkit to allow for home shopping and e-commerce. Downloading of different text fonts and text input from a remote control or keyboard are also supported. The EuroMHEG API has other functions which are a development of the original open standard MHEG-5 API, and as such is regarded as a migratory step from MHEG-5 toward the DVB-MHP. This is useful for broadcasters and network operators who wish to launch a DTV service prior to the standardization of the DVB-MHP.

9.6.4 Data Carousels

Data carousels are of interest to DTV service providers and network operators. A carousel is defined as a rotating magazine, for example slides in a projector, or luggage on a rotating conveyor belt. In broadcasting carousels have been used for a long period of time in conjunction with analog services such as teletext, where the contents of the carousel are cyclically repeated and the repetition rate is quite low. The data is broadcast to some defined playout, and the simplest playout is a carousel or rotation of information. If a receiver wants to access a particular module it simply awaits the next time the data from that module is broadcast. These simple data broadcasting techniques are popular with the public, despite the basic nature of the service provided.

It is expected that DTV carousels will be able to improve on these carousels and approach near multimedia presentation, through the better delivery of text, support for audio-visual clips and bitmaps, easier navigation, better graphics, and better scheduling.

Data broadcast according to the data carousel specification of the MPEG-2 DSM-CC is transmitted in a data storage media command and control (DSMCC) data carousel. The DVB data broadcast specification for data carousels supports data broadcast services, which require the periodic transmission of data modules through DVB compliant networks. Again, the application decoder must await the transmission of a particular module to access the data contained within that module.

For the support of interactive services DVB has adopted another part of the DSM-CC specification, known as the object carousel, which provides the facility to transmit structured groups of objects from a broadcast server to specific receivers. This provides enhanced capability as compared to the above mentioned data carousel [6].

9.6.5 Resident and Interactive Applications

A part of the middleware is often referred to as an application, and consists of software code which is used to allow a user to interact with various services and products. If the application is resident, that is, resides within the set top box, then it can be loaded into memory when the set top box is initially switched on. It may allow for tuning of the STB to various channels, set-up configuring, and also manage any interactions between the user and the service provider. Alternatively interactive applications require the user to authenticate the requested service or product from the provider before downloading, and hence control of access to the application remains with the service provider.

9.6.5.1 Electronic Program Guide (EPG)

The electronic program guide has been available in limited form with analog television, however it is expected to change dramatically with DTV. With DTV it allows the consumer to navigate through the various services and programs offered by the network operator or service provider from a graphical user interface (GUI). The EPG is regarded as a very powerful tool in guiding the consumer through the vast amount of services and programs offered in a DTV platform. The EPG will be able to present choices to the consumer based not only on a channel by channel basis (as is the case with analog television), but also by categorization of program content into different interest groups. It enables the consumer to create a viewing schedule based on content. It contains some of the information contained within a traditional paper based television listing service.

The EPG will use service information (SI) as the basis of most of the information that it will display (see Chapter 5). However, it will need to be collated and presented to the consumer in a responsible and coherent manner. Competition issues may arise on a platform between different service providers if undue prominence of a particular service is shown over others. Therefore some degree of regulation in the provision of the EPG is expected.

Some of the features that will be available from an EPG include program guides for a number of days in advance. Program browsers will display the current and next programs on all available services and allow for immediate switching of channels. These browsers will be available for both audio and video services. It will also allow for notification of events on other channels while the consumer is viewing a completely separate channel. Interactive services will be accessed through the EPG, for instance access to the Internet, email, e-commerce, and games, to mention but a few. The power of the EPG is realized when the complete platform is available through a single EPG.

References

[1]   ETR 290, "Digital Video Broadcasting (DVB); Measurement Guidelines for DVB Systems," May 1997.

[2]   DVB Document TM1748, Draft MG 119 Rev. 1,"DVB-T Measurement Guidelines," September, 1996.

[3]   ITU-T Recommendation 0.151: "Error Performance Measuring Equipment Operating at the Primary Rate and Above."

[4]   Nokes, C., "Bit Error Rates for DVB-T Signal," *Digital News*, Digital TV Group, No. 7, February 1999, p. 18.

[5]   AGITS/WG7 Document, "Digital Television—A Preliminary Guide," 1999, http:/www.teltec.dcu.ie/agitswg7/dtv/

[6]   Horst, H., "DVB Data Broadcasting: Building the Info Highway," *World Broadcast News*, Special Supplement, November 1998, pp. 16–21.

CHAPTER

10

Contents

Single Frequency Networks and Multifrequency Networks

10.1 Introduction to Single Frequency Networks

The possibility of establishing single frequency networks (SFN) is a direct consequence of choosing coded orthogonal frequency division multiplexing (COFDM) as a modulation method. This possibility is only available with the DVB-T standard, and it is regarded as a major benefit of the technology by national radio frequency regulatory bodies and broadcasters alike throughout the world. This is because there is a shortage of suitable UHF and VHF broadcast channels for digital television services (see Figure 10.1). As the new digital channels are brought into service the frequencies of these channels must be co-ordinated internationally (see Section 8.2.2) to protect other digital services and existing analog television services. By using single frequency networks it is hoped by many that it will be

No. of transmitters
per UHF channel

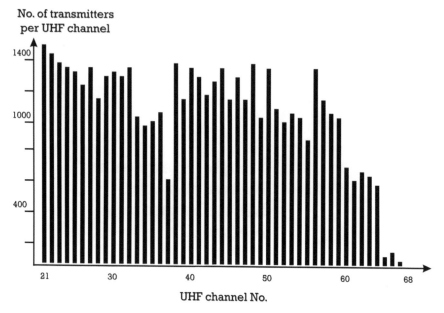

Figure 10.1   Approximate usage of UHF spectrum in Europe.

possible to overcome the restrictions on channel assignments and bring more digital services on air. However, it is accepted that using single frequency networks poses problems for broadcasters who are mandated to provide other service information, known as SI other (see Section 5.5.3 about other services sharing the same mainstation site). It is also constraining on the data broadcasting capacity of the broadcaster as all transmitters must broadcast the exact same information at all times within the single frequency network. This limits the potential capacity of wireless systems for data service delivery to consumers. In this chapter multifrequency digital television networks will also be examined, as these are being widely deployed at present where suitable spectrum can be found for the new services.

A single frequency network is one where all the transmitters within the network are operating on the same RF channel and where all the transmitters are modulated synchronously with the same data signal (see Figure 10.2 for a conceptual view of such a network). The network is only possible to establish due to the usage of COFDM technology as described in Chapter 7. It is also dependent on the correct synchronization of the transmitters in the network, by distributing a stable frequency

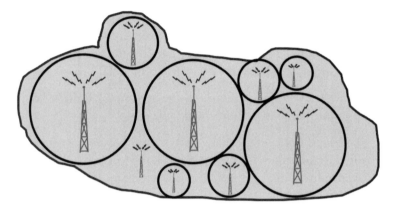

Figure 10.2 The SFN conceptual view.

source and timing reference pulse to the main stations. It is also important that all the transmissions are bit identical with regard to the actual data signal, and this has implications as we will see for MPEG-2 signal distribution. It relies on the fact that once established, the network can be adjusted to ensure that any signals simultaneously arriving at a receiver will add constructively and without interference. This is achieved by careful selection of the system mode (2k or 8k) and by varying the guard interval duration to match the expected network transmission delays. The reader is recommended to refer to Chapter 7 for a more detailed explanation of COFDM parameters before reading on.

10.2   Types of Single Frequency Networks

Various different types of correlated and uncorrelated SFNs are possible to establish depending on the requirements of the network operator and the regulatory conditions prevailing in the region.

A correlated SFN is one where the program content is identical throughout the network. An uncorrelated SFN is one where there is no correlation between signals from different transmitters, either in terms of content, or synchronization.

For the purposes of this chapter the more sophisticated correlated SFN only will be treated, as the uncorrelated network can be thought of in terms of the frequency re-use of the network. The more important SFN types include:

1. National SFN where there are many high power transmitters with large transmitter separation. These typically will cover a large area like a complete region or country, and would be most applicable to a national broadcast service, without regional opt-outs.

2. Regional SFN where there are one or more high power transmitters with large transmitter separation.

3. Localized SFN where main station high power transmitters may operate on different channels but have localized clusters of SFN transmitters in close proximity to each main station.

4. SFN gap fillers which are low power co-channel transmitters fed with an off-air signal from the main station, and which are typically used to fill in the coverage gaps.

5. Hybrid localized SFN where each high power main station operates on a different channel but feeds transposers all operating on a common channel. These transposers may or may not be synchronously modulated.

10.3 Self Interference of Single Frequency Networks

As mentioned in the introduction it is extremely important to select the most appropriate COFDM operating mode when designing an SFN. The two most important parameters are the choice of the guard interval (GI) and the system operating mode (2k/8k) for the SFN. In Table 10.1 below is shown the different guard interval durations in microseconds for different modes and guard intervals. Immediately obvious is that the 8k mode gives a factor of 4 increase in guard interval duration for the same guard interval fraction over that of the 2k mode. This happens without a loss in data capacity and is one of the reasons for the popularity of the 8k mode.

The self-interference of the SFN arises due to the transmitter separation distance exceeding that which the guard interval can accommodate. If signals from a number of transmitters which are distant from each other arrive at a receiver the signals may be delayed in time relative to each other due in part to the physical separation of the transmitters. If this distance is large enough it can create a delay, which exceeds the

Table 10.1
Guard Interval Fractions and Time Durations

Guard interval fraction	2k mode	8k mode
$\frac{1}{4}$	$56\,\mu s$	$224\,\mu s$
$\frac{1}{8}$	$28\,\mu s$	$112\,\mu s$
$\frac{1}{16}$	$14\,\mu s$	$56\,\mu s$
$\frac{1}{32}$	$7\,\mu s$	$28\,\mu s$

guard interval duration chosen for the SFN. If this condition occurs the delayed signal will act like noise and cause interference to the receiver. The magnitude of the interfering signal will be seen to vary with time and also show a dependence on localized propagation conditions. Network designers usually specify as a rule of thumb, that the selected guard interval should at least match the expected delay from the physical separation of adjacent transmitters in the network [1].

10.3.1   Delay Spread

The delay spread is the difference in the time of arrival of two signals (carrying the same information) at a receiver. It is independent of the transmission frequency used. In Table 10.2 guard interval durations for all combinations of guard interval and operating mode (2k/8k) are shown. This also corresponds to the maximum delay spread allowable when implementing a SFN. It also shows the resulting maximum physical separation possible for two adjacent transmitters, according to the simple rule of thumb mentioned above for selecting appropriate SFN guard intervals.

However it is more likely that a receiver will be positioned within a service area of a wanted transmitter (T_{x1}) and will receive an unwanted signal component from a transmitter (T_{x2}) in an adjacent service area as shown in Figure 10.3. In this case the delay spread is due to the difference in path lengths to the receiver, and not just the separation of the transmitters.

Path length difference $D = d2 - d1$ [km].
Assuming $d2 > d1$.
$$\text{Delay spread } T = \frac{(d2 - d1) \times 10}{3 \times 10^8} \text{ [sec].}$$

Table 10.2
Maximum Delay Spread and Transmitter Separation for Different Guard Intervals

Guard interval fraction	2k mode maximum delay	2k mode maximum transmitter separation	8k mode maximum delay	8k mode maximum transmitter separation
$\frac{1}{4}$	56 μs	16.8 km	224 μs	67.2 km
$\frac{1}{8}$	28 μs	8.4 km	112 μs	33.6 km
$\frac{1}{16}$	14 μs	4.2 km	56 μs	16.8 km
$\frac{1}{32}$	7 μs	2.1 km	28 μs	8.4 km

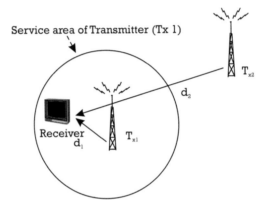

Figure 10.3 Delay spread due to two transmitters.

As was shown in Table 10.2 the maximum guard interval of 1/4 in the 8k system will allow for a maximum path length difference D of approximately 67.2 km, which corresponds to 224 μsec. Thus if the path length D, exceeds this distance then interference may be expected. However, fortunately the interference is characteristic of noise and will generally tend to be of a lower magnitude than the wanted signal. Also a fixed receive aerial pointed toward the wanted transmitter will have some degree of signal discrimination due to orientation and further reduce the unwanted signal somewhat. It is generally accepted that in most reception conditions unwanted interference from adjacent service areas within a SFN will not cause destructive interference. However with portable and mobile reception using a whip-type omnidirectional receive aerial, no antenna discrimination will be afforded and interference is more likely.

10.4   SFN Performance With 0 dB Echoes

Most experimental work with Single Frequency Networks has involved the establishment of networks with two or three co-channel transmitters. With this simple arrangement it has also been possible to vary the time delays between the two signals arriving at a reference receiver. It has also been possible to ensure that the magnitude of the signals is varied to allow the performance of COFDM single frequency networks to be tested under the most extreme conditions. A common parameter tested is the SFN performance in the presence of equal level signals, or 0 dB echoes. This is performed in conjunction with time delays that are at the very extremes of the guard interval duration specified for (DVB-T) SFN compliant operation. These types of extreme echoes are not expected to be found in practice, however they allow benchmarking of different guard interval performances to aid in the selection of a suitable guard interval duration [2].

It is worth pointing out that the benefit of using a single frequency network is not the fact that multiple signals are expected to be received simultaneously by a receiver. A well-designed network will only have fringe areas where it may be possible to have some small degree of over-spillage from service area to service area. The benefit of an SFN is in spectrum efficiency, that is, re-using the same channel over and over again, throughout the network.

It has been found experimentally that when two signals of the same magnitude are incident at a receiver (termed a "0 dB echo") from two transmitters in an SFN, the receive signal power is doubled (carrier power increases by 3 dB). However the required carrier to noise ratio for error free reception is also doubled (C/N ratio increases by 3 dB). This fact has been reported by some commentators to indicate that an SFN will not work with 0 dB echoes, however this is incorrect as it does! The fact that the required C/N ratio increases, is equally offset by the corresponding increase in received signal power (carrier) due to the 0 dB echo. So 0 dB echoes can be tolerated in an SFN, but they will not result in network gain as the required C/N ratio will also increase.

However it is statistically unlikely that a fixed receiver would ever receive equal magnitude signals due to network planning. The carrier to noise penalty, mentioned earlier, for 0 dB echoes in a single frequency network, decreases as the difference in relative magnitude of the two signals increases. For example two signals received with a 10 dB magnitude difference would cause only a few tenths of a dB increase in carrier to noise requirement over that required for each alone in the network. Typical digital single frequency networks will be designed for approximately 23 dB relative difference in magnitude, between the wanted signal in a service area and any co-channel incidentally received signal (termed protection ratio). Therefore it is unlikely that 0 dB echoes will ever be encountered within the service area of a main station transmitter in an SFN.

This 0 dB echo performance at first sight tends to obscure the main benefit of using a SFN, that is a spectrum efficiency. As was mentioned in Section 10.1 this is achieved when all transmitters in the network use the same broadcast channel for the same digital television service, with bit identical transmissions and are time and frequency locked together. Then it is possible to use only this one channel for complete regional or national coverage. This is not possible with analog television transmission systems and is a major technical innovation.

10.5 Single Frequency Network Synchronization

Most digital television (DTV) broadcasting networks will use a digital primary distribution network to feed the MPEG-2 transport streams from television studios and playout centers to remote remultiplexing sites and transmitters. Digital telecommunication networks are the most common distribution channels used. These will include microwave link systems, employing PDH multiplexing and/or SDH multiplexing, and digital microwave radio systems. Other choices for distribution include satellite and optical fiber [3]. ATM networks are becoming popular for distribution as they offer increased flexibility in network usage and efficiency in network management over other architectures, but introduce an overhead which has limited their usage to date. Standards have been developed for interfacing MPEG-2 signals with these network types at all the common network operating bit rates. For all types of distribution network, careful timing control is needed in order to minimize the amount of timing jitter added to the MPEG-2 transport streams.

10.5.1 Synchronization Constraints and MIPs

As mentioned earlier in Section 10.1, for an SFN to work correctly every transmitter must broadcast the same bit of information, at the same time, and on the same channel. This results in 3 synchronization constraints being imposed on the SFN network and careful design is needed to

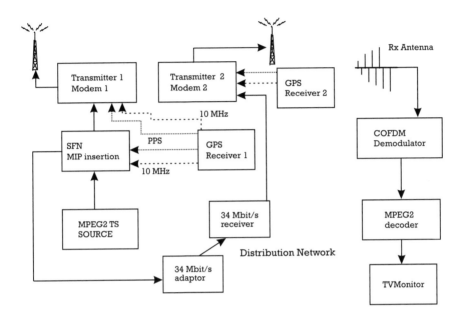

Figure 10.4   Two transmitter SFN synchronization example.

ensure all 3 are met simultaneously. An example of a two transmitter SFN distribution scheme including synchronization is shown in Figure 10.4, with the timing and frequency signals as well as the MPEG-2 transport stream MIP (Mega Frame Initialization Packet) insertion, according to ETSI specifications (TS 101 191) for SFN synchronization [4].

10.5.2   Time Synchronization

This requires the distribution of an external stable timing signal to all COFDM modulators (and/or transmitters) in the network. This pulse must be at least as accurate as ±1 μsec. Also the MPEG-2 encoders and multiplexers may need access to this timing reference. The most common way to supply this is from a 1 pulse per second (PPS) TTL (transistor transistor logic) timing signal, which is available from a global positioning satellite system receiver (GPS).

10.5.3   Frequency Synchronization

Most high power terrestrial analog transmitters are already precision frequency offset and have been using a reference 10 MHz (50 Ω source impedance) source from a GPS receiver for many years. As a result it is likely that transmission sites which are currently broadcasting analog services will have a stable 10 MHz GPS reference. The same GPS reference is adequate for frequency locking the COFDM modulators, and MPEG-2 equipment. However, most older GPS receivers that are used by analog transmitters may need extra timing and frequency outputs, as each modulator requires a separate time and frequency reference.

10.5.4 Bit Synchronization Using the Megaframe and MIP

In order to ensure that all transmitters are broadcasting the same bit of information from the MPEG-2 transport stream (TS) at the correct time, it is necessary to add a timestamp to the MPEG-2 TS. This is done by using a megaframe which is defined as a set of eight 8k frames or thirty-two 2k frames. The time duration of a megaframe ranges from approximately 500–610 msec, and this allows all transmitters to synchronize within approximately 1 sec.

The megaframe contains exactly 1 megaframe initialization packet, denoted a MIP, which contains a timestamp indicating the time at which the megaframe should be broadcast. This is then compared to the universal time and frequency reference available from the local GPS receiver. From these the exact time can be set to broadcast that particular piece of information.

10.6 Distribution Delays

When establishing an SFN it is important to compensate for any differences in time between the different transmitter broadcasts due to the inevitable propagation and processing delays within the network. This equalization is necessary to ensure that all broadcasts are bit identical at the same instant. All distribution networks add a variable amount of delay to a signal as the signal is electronically processed and passed around the network from distant locations to location. In particular digital networks have characteristic long delays that are very well known to those involved in outside broadcasting (OB). These delays can be calculated in advance and compensated for, by adding a maximum delay figure to the MIP timestamp. This maximum delay figure is an amount of time (typically < 100 μseconds), which is chosen to exceed the maximum delay that could be expected in distributing a signal to the most remote transmitter in the network. While the signal is travelling to this remote site all other transmitters are made to wait for this maximum delay time to ensure all transmitters will be able to broadcast the same bit of information at the same time. This allows the network planner to compensate for, and equalize distribution delays in the network. This feature ensures that all broadcasts in the network can be time and bit synchronized. The modulators within the transmitters simply store the MPEG-2 TS in buffers until the predefined moment when all transmitters are ready to broadcast the same bit of information.

10.7 Network Gain

Single frequency networks are an exciting and promising technology, primarily because of spectrum efficiency, however they can also be power efficient if planned correctly. Extensive theoretical studies have been made both for DTV and DAB (Digital Audio Broadcasting) Single Frequency Networks, using computer models [5], to optimize SFN configuration and establish minimum transmitter powers [6]. From these studies a resulting idealized network lattice of equilateral triangles would emerge as shown in Figure 10.5, with each transmitter at a lattice point and having an omni-directional antenna pattern. If then an omnidirectional or whip type antenna were used for reception in a SFN it would receive equally, signals from all directions. This time offset composite signal would have signal components from all adjacent transmitter sites, provided they too were broadcasting towards the receiver (i.e., have an omnidirectional antenna pattern).

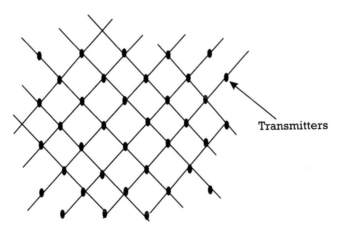

Figure 10.5 Idealized SFN transmitter lattice.

In conventionally planned networks a common way to ensure service availability at the majority of receive locations within a transmitter service area is to include a generous fade margin in the link budget. This results in higher power transmitters being used to accommodate potential fades in signal strength.

In the above planned SFN, any whip receive antenna will be gathering signal components with field strength magnitudes, which are not strongly correlated, from many adjacent transmitters. If one signal should fade then the receiver may have enough compensating field strength from another transmitter to smooth the effect of the fade and prevent complete signal loss. Thus a single frequency network may operate at reduced transmitter powers and enjoy network gain over conventionally planned networks.

This equidistant transmitter separation is not a feature of conventionally planned networks, typically with directive broadcast antenna systems. Also it is more difficult for network operators to obtain the number of sites (as required in Figure 10.5) in the specified locations required for a properly planned SFN, then it is to re-use existing sites. As a result network gain with conventionally planned networks is only expected in fringe areas of reception, between two adjacent service areas.

10.8 Multifrequency Networks

Multifrequency networks (MFN) are more commonly used for DTV broadcasting by network operators than the just described single frequency network arrangement. In this case the modulating signal could be any of the previously described modulations from Chapters 6 and 7. This includes Vestigial sideband modulation (VSB), COFDM, and QAM signals. These conventionally planned networks have transmitters that carry independent program signals and use different RF channels at different locations. The planning rules used to assign channels for a particular multiplex throughout a region or country are similar to those used in conventional analog television planning, with the overall intention of minimizing interference to other RF spectrum users. The channels assigned are co-ordinated between jurisdictions and network operators under the management of bodies such as CEPT. With digital television signals the level of interference which can be tolerated is usually much higher than that tolerable using analog television modulations, and depends on the modulation type chosen and the error correction code chosen. Quite often digital multiplexes are interleaved between existing analog transmissions in order to simplify the receive aerial installation, as described in Chapter 8.

10.9   Protection Ratios

Protection ratios are used extensively in planning networks where frequency reuse can be expected. It is the same definition that is used whether a digital or analog network is in question. However the value of the protection ratio will depend greatly upon the signal type employed to modulate the data onto the RF carrier, and whether a frequency offset is used. It is also dependent on the level of coding employed to provide error correction. The protection ratio is defined as the minimum ratio of the (power of the) wanted signal to the (power of the) unwanted signal measured at a receiver. This topic will be covered later in Chapter 11.

Protection ratios can be defined for co-channel interference and also for adjacent channel interference. Co-channel protection ratios for DTV are very dependent on the system employed to modulate both the wanted and interfering signals, and also the relative power levels between them. Adjacent channel protection ratios are also dependent on the level of filtering employed in the transmission system to limit out of band emissions, and also on the selectivity of the tuner front end in the receiver. A high selectivity channel filter will reduce the adjacent channel levels reaching the demodulator, and therefore reduce the level of interference.

Protection ratios are calculated using some measure of objective signal degradation. For digital signals this is usually the bit error rate (see Chapter 9). There are different levels of acceptability for the degradation of the wanted signal depending on the duration of the degradation. So-called *continuous* protection ratios apply for at least 50% of the time, whereas *tropospheric* protection ratios apply for only 1% of time. Tropospheric protection ratios will be dealt with in Section 11.6.1 when discussing interference from distant co-channel transmitters in a single frequency network.

The protection ratio depends strongly on the nature of the interfering or unwanted signal. Various ratios are proposed for digital signal to digital signal interference, analog to digital interference, digital to analog interference, and the well known analog to analog interference. Recent field measurements indicate for COFDM systems that the code rate is the most important parameter in lowering the required protection ratio for a DTV signal interfered with by either a co-channel analog TV signal or DTV signal [7].

10.9.1 Frequency Resource

The frequency resource of a network is the number of channels that are required to provide a desired signal at any location within the network. For an SFN it is unity, but it is a lot higher for an MFN. For an MFN it is dependent on the protection ratio required for the modulating signal because if the DTV system has a low protection ratio this means that it can tolerate quite a high level of co-channel interference. As a result the same channel can be re-used in the network more often. However as we have seen in earlier chapters the classic cliff effect for failure of digital signals means that any DTV system will tolerate interference until a threshold level is reached, and after that level abrupt and sudden failure will happen. This means that designers must allow for this failure mechanism and add a margin of safety to allow for occasional occurrences such as signal fades due to atmospheric propagation.

10.9.2 Assigning Channels for MFNs

Research carried out in the early 1990s as part of the SPECTRE project had identified that it would be possible to operate digital terrestrial services in channels adjacent to the existing analog television channels, the so-called taboo channels [8]. In Chapter 8, an example was shown of a conventional analog PAL I mainstation with four channels assigned within the same channel group. In Figure 8.1 it was illustrated that many channels within the group were not available in the past for use due to restrictions placed by receivers and transmission combiners.

However in the planning of a DTV system from the same site, these taboo channels, which were not available in the past for analog services, can be used for DTV transmission. This is desirable because these channels could then be received by a consumer without having to purchase a new receive aerial. The analog channels already in use are UHF channels 40, 43, 46, and 50. This leaves channels 41, 42, 44, 45, 47, 48, and 49 within the same channel group as potential DTV channels. The actual allocation of the channels within this group would need thorough planning and coverage prediction analysis using the predicted ERP (emitted radiated power) levels for each service. This would take into account the likelihood of interference to/from any co-channel broadcasts at nearby sites, both inside and outside the network.

10.9.3 MFN Offsets

Frequency off-sets are used extensively in conventional analog television networks, and in particular precision offsetting is a technique which is used to dramatically lower the required protection ratio when two co-channel transmitters are likely to interfere with each other [9]. The two transmitters typically use a GPS frequency source to derive a 10 MHz reference source which is used to ensure an exact fractional line offset is applied to the two transmitters at all times. This relative offset is defined typically in terms of fractions of the offset frequency.

Offsetting has been planned for in DTV systems, but not in the same manner as for analog offsets. It is not used in DTV to reduce interference but to facilitate easier adjacent channel combining. A commonly accepted method is to make provision for a maximum allowable offset value, and a typical maximum value quoted is ±200 kHz. In the United Kingdom offsets of ±166.666 kHz (1/6 MHz) are used when adjacent channel combining of analog and digital services is performed. It makes the channel combiner easier to realise for the network operator and it facilitates easier reception for tuners in set top boxes. The problem originated due to the difficulty of combining a DTV signal with a lower adjacent analog PAL I signal, which has a NICAM signal at the channel edge. There was a lot of debate over the issue of offsets and the potential to degrade the front end (phase noise) performance of the receiver, however this value of offset is in use without degradation problems being reported.

10.10 Nonsynchronous Operation

As was mentioned earlier in this chapter, for a single frequency network to operate correctly all of the broadcasts from all of the transmitters must be synchronous. This means in effect that all broadcasts must contain the same data, they must happen at the same time and on the same frequency. However these constraints do not apply to an MFN, and as a result the network is much easier to configure and is also more flexible in operation. There is no need to co-ordinate the emissions from different mainstation sites. Delays in signal distribution throughout the network do not need to be accommodated when timing the broadcast emissions. It is possible to allow for regionality of service, where a local or regional service is added to the multiplex. Service information (SI) and especially SI other information can be added easily at any mainstation site. The regional services could then employ a regionalized SFN system if this is desirable within the overall MFN.

10.11 ERP Requirements

Terrestrial propagation is characterized by a significant location variation of receive signal field strength at times. This variation is due at times to enhanced propagation and at other times to signal fades. As a result only location coverage probabilities are quoted for planning purposes. With analog television graceful degradation was noticed as the signal faded, especially on the boundaries of the service area. This meant that if a signal faded in a location, it was likely to fade gracefully and reduce in quality only. However digital signals fade more dramatically and as a result designers must ensure that greater location coverage probabilities are provided for. This results in higher ERP levels for DTV signals to account for poor coverage in fringe areas on the edge of the service area. This is referred to as excess power. However if the service areas of adjacent transmitters in an MFN are designed to overlap at the fringe areas, then it is likely that a receiver could select the strongest signal level from these adjacent transmitters at any time. This is believed by some broadcasters to allow a reduction in transmitter excess power requirements, however it would require the consumer to retune their receiver to the adjacent service area, and implies the use of an omnidirectional antenna. This is not acceptable to many broadcasters and as a result higher power will be required to ensure that a higher percentage of locations are covered within the mainstation service area.

10.12 Modifications to Upper Adjacent Analog Transmitters

In the United Kingdom, Ireland, and some countries in Africa, and elsewhere, system PAL I is used for analog television broadcasting. This system, which employs vestigial sideband shaping, has significant out of band or lower adjacent emissions. As a result significant interference to a lower adjacent DTV transmitter could be expected at the top of the DTV channel from an upper adjacent analog signal. To overcome this problem in the UK it was proposed to change the analog transmitters from the original PAL system I to a modified PAL system B/G vestigial sideband which would reduce the lower adjacent emissions [10]. It resulted in only a small decrease in analog television receiver performance and allowed lower adjacent channel operation for the recently launched DTV service.

References

[1]   "ACTS Validate Project: Implementation Guideline for DVB-T," *DVB Document TM1825* Rev. 2, April, 1997.

[2]   O'Leary, S., "Field Trials of an MPEG-2 Distributed Single Frequency Network," *IEEE Trans. On Broadcasting*, Vol. 44, No. 2, June, 1998, pp. 194–208.

[3]   CEC Deliverable Number AC106/BBC/DR/021/P/a1, June, 1998.

[4]   ETS 101 191, V1.1.1 (1997-04), DVB : DVB Mega-frame for SFN Synchronization.

[5]   Weck, C., "Coverage Aspects of Digital Terrestrial Broadcasting," *EBU Technical Review*, No. 270, Winter 1996, pp. 19–30.

[6]   Rogers, P., "Sound Broadcasting - AM to FM to DAB" *Institution of Engineers of Ireland, Callan Memorial Lecture*, February 1994.

[7]   O' Leary, S., "Digital/Analogue Co-channel Protection Ratio Field Measurements," *IEEE Trans. On Broadcasting*, Vol. 44, No. 4, December 1998, pp. 540–546.

[8]   Morton, C., and G. Verity, "RF Engineering for Digital Terrestrial TV," *IBC*, Amsterdam, September 1998, pp. 300–305.

[9]   CCIR Rec. 655, "Radio-Frequency Protection Ratios for AM Vestigial Sideband Television Systems," Report 481, Rec. 655, 1986, pp. 233–249

[10] Marshall, P., "Preparation for Digital Terrestrial Television in the U.K.," *IBC*, Amsterdam, September, 1998, pp. 266–271.

Radio Frequency Considerations for Digital Terrestrial Television Services

11.1 Introduction

Terrestrial broadcasting of television generally involves the high power transmission of signals mainly in the ultra high frequency (UHF) band, which is approximately from 480–860 MHz. It is also possible to broadcast signals in the very high frequency (VHF) band, which is approximately 174–230 MHz for band III of the VHF block. At these terrestrial frequencies signals exhibit good propagation, and quite considerable areas of terrain can be covered from a single transmitter site. There is generally no need for any pre-amplification of the signal between the receive antenna and the television receiver. Portable reception using a whip-type antenna is often possible.

These features of portable reception with terrestrial networks are not possible with other transmission media such as, microwave multipoint distribution systems

(MMDS), satellite, and cable networks. These systems need more complex distribution networks and receiver installations to provide similar coverage levels. Portable and mobile reception of radio and television services can only be provided using terrestrial networks. Cable and MMDS networks require fixed installations, where the receiver is connected to either a cable outlet socket or to a microwave antenna and block down conversion unit.

It is generally accepted that the cost of establishing a terrestrial broadcast network is orders of magnitude fewer than that of the other transmission media just mentioned. This is because of the high terrain coverage of UHF and VHF broadcasts. Less sites are required to cover a geographical area, and no cabling is required to connect the broadcasting signal to domestic receivers.

Governments tend to want to have terrestrial services provided for, because of the highly regulated nature of these broadcasts, and because of the widespread penetration of terrestrial services to the general population. Finally, regional and local broadcast services can be easily provided using a terrestrial based transmission network. For these reasons terrestrial digital broadcasts of radio and television are highly desirable and subject to regulation.

With the advent of digital technologies, terrestrial broadcasters can now offer a wider variety of services to the public, than was hereto possible with analog technology. These services include multichannel television, high quality audio and radio services, data broadcasting, text-based services, email, Internet access, and e-commerce. However, all terrestrial broadcasts have the potential to propagate into other jurisdictions, and cross borders. Since the terrestrial spectrum is very attractive to broadcasters and is of a finite bandwidth, a high degree of spectrum re-use is performed throughout the world. There are many other electromagnetic radiation users and as a result of the popularity of wireless technologies the radio frequency (RF) spectrum is in high demand and used all over the world. The terrestrial spectrum must be managed, just as all scarce resources are managed and protected. As a result, many international and national bodies have been established to make and enforce agreements concerning frequency usage. For any high power terrestrial transmitter to be licensed it must conform to strict regulations concerning the emitted radiated power (ERP) level and bearing of the signal (see Chapter 8). The signal itself must be classified, and interference from the transmitter to other RF spectrum users eliminated. This chapter will discuss some of the coverage aspects of digital broadcasting. It will focus on DTV signals, and in particular discuss the typical protection ratios that are used with the DVB-T signal. It will outline some of the requirements needed to provide for fixed, portable, and mobile reception of DVB-T signals.

11.2 Requirements for Fixed, Portable, and Mobile Reception

11.2.1 Introduction to Digital Coverage

All digital television signals have been developed to provide at least fixed reception. However, some types of digital signals are better than others for mobile and portable reception. This is because some digital signals employ powerful algorithms to ensure that any artifacts generated by portable and mobile reception conditions are eliminated. In general, problems associated with poor fixed reception areas are also found with portable and mobile reception of DTV signals. So advances in the technology of combating poor reception in the presence of, for example, multipath signals, will not only improve fixed reception but will aid toward better portable and mobile reception, generally. It is a well known disadvantage of digital television that the coverage characteristic is that of a "cliff," where a service area can have a rapid transition between signal availability and possibly no reception at all, very rapidly. As a result careful planning is required to ensure that the benefits of DTV are brought to the consumer.

11.2.2 Field Strength and Carrier-to-Noise Ratio

In earlier chapters it was shown that the carrier-to-noise ratio is a very important parameter used to measure the quality of a received signal. It was also shown that different types of signals require larger or smaller carrier-to-noise ratios, to ensure that the digital signal bit error rate is sufficiently low. If the carrier-to-noise ratio is not high enough then the bit error rate will exceed the level that the receiver can correct. This will result in residual errors being displayed on the television receiver, and if the bit error rate increases still further due to reducing the carrier-to-noise ratio, then total signal loss will occur (see Chapter 9). In the subsequent sections the receive signal field strength will be used as a coverage assessment parameter. It should be noted that these field strength requirements are derived using the minimum required carrier-to-noise ratio of the particular modulation chosen to broadcast the DTV signal.

The intensity of an electromagnetic wave or field strength is generally stated in terms of volts per meter of space. This is the strength of the electric lines of force at the receiving aerial and it is at right angles to the direction of propagation of the wave. Field strengths normally range from low values of 1 microvolt per meter to 10–100 millivolts per meter. It is customary to define field strength in terms of a level in decibels relative to a field strength of 1 microvolt per meter. A field strength of 1 mV per meter would then be defined as 60 dBμV/m. For DTV signals it is important to measure the field strength of the entire signal bandwidth, unlike analog television, which measures the peak voltage of the vision carrier only.

11.2.3 Analog Versus Digital Coverage Failure

Analog television can be characterized as having a soft coverage area transition at the edge of the service area. This is because at the edge of an analog terrestrial broadcast zone, where the transmitter no longer provides the required field strength for clear and error free reception of signal, there is usually a graceful degradation of signal quality. The picture gradually gains more and more noise until it finally becomes unwatchable. However, despite the fact that digital signals will suffer a similar graceful reduction in field strength as one moves out of a transmitter coverage area, the perceived failure mechanism is quite different. This is because of the powerful error correction techniques that DTV signals employ. A DTV receiver can tolerate significant loss of field strength without any displayed error being superimposed on the television screen. However, once a threshold error rate is passed, the receiver completely locks up and fails very quickly. This is a problem for network planners, who with analog television normally had the comfort of zones around any main station site where the signal quality gracefully degraded.

11.2.4 Location Coverage Probability

Network planners use the measured field strength of a signal as a measure of the coverage of a transmitter. The number (or percent probability) of locations within a defined area which achieve the required field strength is used as a criteria to assess the coverage of a transmitter. In Europe and elsewhere, new percent location coverage probability levels have been adopted, because of the DTV coverage failure mechanisms just described in Section 11.2.2.

11.2.4.1   Analog Versus Digital Location Coverage Probability

With analog broadcasts it was acceptable if the transmitter covered 50% of locations within any defined area of a transmitter's service area. If this coverage level was achieved then there was a high probability that the remaining 50% of locations would have an adequate field strength to allow reception of signals to acceptable quality grades. However, it was agreed by network operators and regulators after much simulation and field measurement that a higher location probability would be needed to ensure adequate DTV signal reception. The new minimum location probability is 70% of locations within a small area (the size of which is usually defined to be 100 meters × 100 meters). Some jurisdictions have chosen higher location probabilities of up to 95%, where portable and possibly mobile reception is required.

11.2.4.2   Time Probability in Coverage

Any single location will be deemed covered if it achieves the required carrier-to-noise ratio for that modulation scheme. This is achieved if the location receives an adequate field strength to ensure that the carrier level of the DTV signal is the required ratio above the receiver noise floor (see Section 11.3). These criteria assume that the reception location does not introduce other degradations to the receive signal, such as those generated from signal reflections from mountainous terrain and large buildings.

   If any location has the required field strength for 99% of the time, then that location is assumed to be adequately covered. If the small area location probability coverage of a transmitter is achieved for 99% of the time, then this small area (100 meters^2) is assumed to be adequately covered. This results in terminology of a small area being covered for DTV reception if at least 70% of locations are receiving adequate field strength for at least 99% of the time.

11.2.5   Fixed Reception

Fixed reception is the mechanism of reception used where the receiver is connected to a rooftop-mounted directional aerial. The aerial is assumed to have some degree of antenna gain at the receive channel used, that is, it can concentrate the received signal field strength. This will have the effect of increasing the terminal voltage presented to a domestic type television receiver, over that of an omnidirectional whip-type antenna. The standard height assumed for a fixed antenna receiver is 10 meters. This type of reception is common in rural areas, and in homes with access for a roof top antenna. It is also normally used in suburban areas with closely spaced housing units.

11.2.6 Portable Reception

Portable reception is the mechanism of reception where the receiver is directly connected to a small antenna, which can be mounted close to the receiving unit. Some types of receivers have built in receive antennas, which can be of the telescopic whip type, or more elaborate. Portable reception is a more general form of fixed reception, which also allows for reception via fixed antenna. However portable reception also includes conditions such as outdoor reception and reception in a stationary vehicle. It is normally used in cities, and is suited to apartment dwellings, and closely spaced housing units. If portable reception of a signal is possible, then it follows that fixed reception must be possible. These are two commonly defined classes of portable reception, class A and class B.

11.2.6.1 Class A, Portable Reception

In this type of reception the receiver could be located outdoors at a height of no less than approximately 1.5 meters above ground level. It is also possible to receive a signal indoors, in an upper floor indoor location, with similar field strengths, and as a result this reception is sometimes defined within class A. The loss of signal field strength as the signal penetrates a wall into a building is termed the building penetration loss. It has been found to be greatest if the room is located at ground level, and reduces somewhat with height.

11.2.6.2 Class B, Portable Reception

In this type of reception the receiver is assumed to be located indoors, at ground level. It is assumed that the receiver is located within a room with a window in an external wall and at a height of no less than 1.5m above ground level. The building penetration loss will be highest in this case for any received signals, and as a result the required field strength to provide service under these conditions will be higher than that required for class A portable reception. This is a common type of reception condition in urban areas, with apartments and closely spaced housing units.

11.2.6.3 Mobile Reception

In this type of reception the receiving unit is moving and the only type of practical antenna is an omni directional whip-type antenna, mounted on the vehicle. It requires the greatest field strength to allow error free reception and is most suited to modulations which can eliminate multi-path signals, such as COFDM signals. Hierarchical modulation can be used to provide for this type of reception (see Chapter 7).

11.3 Field Strength Requirements

The minimum required field strength is a parameter used by network planners to design transmission systems that ensure adequate signal field strength is provided to receivers within the coverage area of a transmitter. If class B, portable reception is required, for example, then it is possible to calculate the required receive signal field strength to provide an adequate receiver terminal voltage. The most important parameters in this calculation are the receiver noise figure, the required system carrier-to-noise ratio for this type of digital modulation, the channel bandwidth, and the receiver noise input power. These formulas are shown below in (11.1, 11.2, and 11.3).

BW	The receiver noise bandwidth (equivalent to the channel bandwidth) [Hz].
F	The receiver noise figure (a figure of merit) [dB].
P_n	The receiver noise input power, which is the required receiver input power to overcome noise[dBW].
C/N	The system carrier to noise ratio requirement [dB].
Z	The receiver input impedance [Ω] (typically 75 Ω for domestic receivers).
P_{smin}	The minimum receiver signal input power [dBW].
V	The minimum receiver input voltage [dBV].
k	Boltzmann's constant ($1.38*10^{-23}$ [Ws/K]).
T_φ	290 K [Absolute temperature scale].

$$P_n = F + 10 \log(k * T_\varphi * BW) \tag{11.1}$$

$$P_{smin} = P_n + C/N \tag{11.2}$$

$$V_{smin} = P_{smin} + 120 + 10 \log(Z) \tag{11.3}$$

From (11.1) the receiver noise input power can be readily calculated, using the channel bandwidth and the receiver noise figure. Then using (11.2) the minimum required signal power, which must be input to a receiver to ensure error free DTV reception, can be calculated. Finally this power requirement can be converted into an input terminal voltage using (11.3). It is important to note that the required field strength for error free reception of a DTV signal is chosen as a result of the carrier-to-noise requirement of the particular modulation chosen. With the DVB-T specification, many variants of modulation are available, as was shown in Section 7.7, and each variant has a slightly different carrier-to-noise ratio requirement for error free reception. This allows tailoring of the modulation to meet transmission requirements.

In Table 11.1 below, a comparison is made between the typical field strength requirements for analog PAL I signal reception and that required by a typical 64-QAM COFDM signal. Note that the analog signal location probability is only given for 50% of locations, whereas the DTV signal is defined for higher location probabilities. It is also worth noting that class B portable reception requires considerably high field strengths.

11.4 Protection Ratios for Terrestrial DTV

Protection ratios are widely used in planning terrestrial networks. A radio frequency protection ratio can be defined as the minimum value of the wanted to the unwanted signal ratio. It is usually expressed in decibels at

Table 11.1
Comparison of Field Strength Requirements for Analog and DTV

Signal type	Location probability (%)	Fixed reception (dBφV/m)	Portable reception class A (dBφV/m)	Portable reception class B (dBφV/m)
Analog (PAL I)	50	64	—	—
DTV (COFDM)	70	46	68	75
DTV (COFDM)	95	55	77	84
DTV (COFDM)	97	57	79	87
DTV (COFDM)	99	61	82	90

the receiver input, and is determined under specified conditions such that a specific reception quality is achieved at the receiver-input [1]. It is a term, which is often used when establishing networks, which have some degree of frequency or channel re-use. For DTV systems, the usual reception quality indication is bit error rate. For DVB-T systems a BER of 2×10^{-4} measured after the Viterbi decoder and before the Reed-Solomon decoder is a suitable quality indication (see Section 9.4).

11.4.1 Co-channel Interference

In all networks frequency spectrum is a scarce resource, and it is invariably re-used in different parts of the same network. In these cases the network must provide sufficient protection against co-channel interference (CCI), where two or more transmitters are using the same broadcast channel. In the case of a multifrequency network (MFN), whenever a broadcast channel is re-used there is potential for interference between the two co-channel stations, and the protection ratio becomes an important parameter in this case. For any receive site, the protection ratio is the relative difference (in dB) between the wanted and the unwanted signal. For DTV signals the minimum required carrier-to-noise ratio (C/N) for the modulation scheme sets the minimum protection ratio. This is because for any DTV signal, an interfering co-channel DTV signal appears to be just like a noise signal. However, in a single frequency network co-channel interference only exists if the two DTV signals are delayed in arriving at the receiver by more than the system guard interval.

11.4.2 Adjacent Channel Interference

Adjacent channel interference (ACI) is another problem caused by relatively high adjacent channel power levels. This can cause problems for receivers that do not have high selectivity when filtering out unwanted channels. Often DTV channels are assigned on adjacent channels to existing high power analog television channels in order to allow reception from the same terrestrial rooftop aerial. This can cause problems for network engineers when establishing DTV networks to be broadcast from existing analog television sites. If the relative levels of the DTV and analog signals are too large, interference from one channel into the adjacent may result. Filtering the input signal at the front end of the receiver has been shown to alleviate the problem in field trials, suggesting that improvements in filtering may offer solutions.

11.4.3 Interference Possibilities

There are many possible situations of interference from a signal to another signal and as a result protection ratios have been established for the more common, including:

- Digital television interfered with by digital television;

- Digital television interfered with by analog television;

- Analog television interfered with by digital television;

- Analog television interfered with by analog television (the existing case, before DTV introduction).

A further complexity is introduced when the various existing analog television schemes, such as PAL I, PAL B/G, NTSC, SECAM, and so forth, are considered. However, these have all been investigated and tables of results are available [2, 3]. Careful planning and coordination are required to minimize interference from adjacent channels, and co-channel services.

11.5 Propagation Curves

The propagation of any broadcast signal from a transmitter to a receiver has been shown to be accurately modeled as a statistical process, with various propagation conditions that have to be considered at various distances from the transmitter [4]. The parameters for this statistical process are based on the results of numerous field trials and field measurements, with different types of signals, and across a wide band of transmission frequencies. As a result, most DTV and DAB field strength predictions use the ITU-R propagation curves as a basis [5]. These curves, which were originally used for analog television, have now been validated for digital television signals through many recent field trials.

11.6 Single Frequency Network Considerations

Single frequency networks offer the distinct advantage over multifrequency networks (MFN), of very high spectrum efficiency for services that are to be broadcast everywhere within a defined network. As was seen in Chapter 10, SFNs have disadvantages compared to MFNs if regional differences are envisaged in the content of the multiplex. SFN operation is a key part of the DVB-T specification, and the DAB specification. Many network operators in Europe are using SFN operation to maximize coverage and spectrum efficiency.

In a perfectly synchronized SFN, it is possible for the signals from adjacent transmitters to be collected by a receiver without any interference. This assumes that the relative delays of the signals from each transmitter do not exceed the guard interval. If this happens, then the protection ratio criteria are required to ensure that the wanted signal (from, in this case, the closest transmitter) exceeds the unwanted signal (in this case, any other signal delayed by more than the system guard interval) by at least the required carrier-to-noise ratio of the particular modulation being used.

11.6.1 Interference From Distant Transmitters in an SFN

For large area SFNs, there will be found many co-channel distant transmitters. The signals from these transmitters may always be considered as interferers to other distant transmitter service areas, as the components reaching a distant service area will cause some interference to receivers in that area. This is because the relative delays of these distant signals will exceed the maximum system guard interval (which is approximately $224\,\mu s$, for DTV). Since the transmitters are distant, it is assumed that the relative difference between the receive signals will exceed the carrier to noise ratio, and reception will be possible. The interfering transmitter is assumed only to cause interference when climatic conditions are favorable to propagation. These interfering transmitters can use the ITU-R propagation curves for 1% of the time, which is because they are only expected to cause significant interference at times of exceptional propagation, e.g., high atmospheric pressure.

By tailoring the size of the guard interval to match the maximum expected delay between adjacent transmitters within a network, the potential for interference can be reduced. However, some level of interference from distant transmitters must be expected occasionally, when climatic conditions favor propagation. The flexibility of the DVB-T system allows network operators to tailor the guard interval duration to meet existing network transmitter spacing. However, new sites may be required to reduce interference if omnidirectional antennas are used on receivers, as is planned for DAB receivers and mobile DTV receivers.

References

[1]   CCIR Recommendation 655, "Radio Frequency Protection Ratios for AM Vestigial Sideband Television Systems."

[2]   EBU Planning Document BPN 005 Rev. 6.

[3]   ACTS VALIDATE, "Implementation Guidelines for DVB-T," DVB Document TM1825 Rev. 2.

[4]   Weck, C., "Coverage Aspects of Digital Terrestrial Broadcasting," *EBU Technical Review*, Winter, 1996, pp. 19–30.

[5]   ITU-R Recommendation PN.370-6, "VHF and UHF Propagation Curves for the Frequency Range from 30 MHz to 1,000 MHz, Broadcasting Services."

Digital MMDS Systems

12.1 Introduction

In this chapter a description of traditional analog microwave multipoint distribution systems (MMDS) will be given. Some of the factors that affect the propagation of MMDS signals in the 2–3 GHz frequency range will be explained. The emerging digital transmission standards that are used with MMDS will be described as well as some implementation possibilities.

A microwave multipoint distribution system (MMDS) is also known as a multichannel multipoint distribution service, as a multichannel microwave distribution service, or simply as wireless cable, for reasons that will become clear when the historical development of MMDS is described in the next section.

12.1.1 MMDS Definition and Historical Development

MMDS can be defined as a fixed service system used for the retransmission of programs on a point to multipoint basis. A typical system consists of a number of microwave transmitters, each of which is coupled to a common channel combining system. The output of this microwave combiner feeds a composite signal to a microwave broadcast antenna located at the end of a transmission line feeder at the transmitting site. This antenna may or may not be omnidirectional depending on the service area to be covered, but will be located high up on a suitable transmission structure such as tower or mast. At each of the domestic receiving sites there will be a highly directional microwave antenna, and a low noise block down converter unit that is used to amplify the received signals and frequency shift these signals from the microwave band to the UHF band (see Section 12.4). If the signal is encrypted then a set top box will be used to descramble the program service just before the signal is input to a television receiver.

MMDS is an adaptation of instructional television fixed service (ITFS), which is a television distribution technology that has been used in the United States for many years to broadcast educational material to schools and colleges. ITFS uses frequencies of approximately 2.5 GHz, and has been in operation since the early 1960s. Multipoint distribution system (MDS) and MMDS have grown out of this early technology as an economic way to provide multichannel analog television in urban and suburban areas. MDS was originally used to provide 1 or 2 premium channels to subscribers over the air. MMDS can provide many more channels and has grown considerably in popularity during the 1980s and 1990s throughout the United States, where it is often referred to as wireless cable. It is also deployed widely in South America, and parts of Europe, where it was seen as a useful way to provide multichannel television in unassigned spectrum around 2.5 GHz. It is often regarded as a natural competitor to cable television networks, having lower network installation costs. However it requires a technical installation at the receive site, and is dependent on a line of sight path from the transmitter antenna to the receive antenna. Also with the advent of digital terrestrial television, MMDS itself now has a major competitor with lower network costs than MMDS, simple installations, and better propagation characteristics at the terrestrial UHF and VHF bands.

12.2   Spectrum and Propagation Factors

As was mentioned above MMDS was developed from ITFS and MDS in the United States. ITFS used the frequency band 2.5–2.7 GHz, while MDS used frequencies around 2.1 GHz. Most deployed MMDS networks today also use the 2.5–2.7 GHz band, which is at microwave wavelengths, and the propagation is therefore limited to the radio horizon, unlike VHF and UHF transmission networks. Table 12.1 below summarizes some of the more important factors associated with MMDS propagation and their effects. These factors are also relevant to discussions on electronic news gathering (ENG) circuits, as we shall see later in Chapter 14.

12.2.1 Common MMDS Propagation Problems and Recommendations

Later in Chapter 14, we will see that microwave signals are known to bend as they travel though the air. As the density of the air changes, due to temperature changes, the signal path is altered. Generally signals are bent toward the ground, which has the effect of extending the radio horizon. As a result the MMDS radio horizon is calculated from 4/3 the actual radius of the earth.

Obstructions to the main path between a MMDS transmitter and receiver will block the signal from reception. However there are several useful paths which can reach the receiver. In addition to the main path, these paths are deflected away from the main path and form an envelope around the main path. This envelope is termed the *first Fresnel zone*, and is defined, as all paths with a length equal to the direct path plus ½ the wavelength at the particular frequency. In order to maximize the signal strength received most designers recommend that a clearance of approximately 0.6 of the first Fresnel zone be maintained around the main signal path between the transmitter and receiver [1].

Table 12.1
MMDS Propagation Factors

Factor	Effect
Terrain shadowing by hills	Very high, 25 dB or more signal attenuation behind hills
Trees and foliage shadowing	Significant, 3–12 dB winter to summer variation, if installation occluded by foliage
Structural shadowing	Significant shadows behind metal structures
Rain fading, snow fading	Low effect; A 2 in/hr rainfall rate will cause 0.02 dB/mile attenuation
Ionospheric reflection	Negligible
Free space propagation	Very high path loss (see Section 12.3.1)
Multipath reflections	Low probability due to the usage of high gain, and narrow beamwidth receive antennas

A famous anomaly caused by reflections is the phenomenon known as ducting, which often occurs around and over large bodies of water (see Section 14.6). Different layers of air are formed over the water that cause the MMDS signal to get trapped within a layer and propagate much further than the radio horizon. Another anomaly is tidal fading, an effect that is linked to long sea paths that have reflections over the sea surface. As the tide rises and falls, the received signal level can be observed to suffer a corresponding change.

12.3   Free Space Attenuation

Free space propagation is a fundamental reference for radio frequency engineering. The attenuation that a signal is subjected to while propagating in free space is a very important parameter for all users of RF spectrum. Atmospheric propagation is generally assumed to approximate free space propagation for attenuation calculations, and as a result free space path loss is used when calculating signal path loss for MMDS networks. In order to transmit information all users of RF spectrum need to know how much power is required to get a signal from a transmit point to another receive point. There are different methods used to calculate the free space attenuation of a RF signal, depending on the nature of the transmission. For instance a broadcast signal is generally regarded as a point to multi-point (or area) link. This is because there is generally one transmitter that is serving many randomly distributed receivers [2]. It should be noted that free space loss (FSL) is often termed simply the *path loss*.

12.3.1   Free Space Loss of MMDS

As was indicated in Table 12.1 above, MMDS suffers from a high free space loss because of the high transmission frequency which is used to broadcast signals. In a similar manner to microwave links it is highly dependent on a clear line of sight from the transmitter antenna to the receive antenna.

The method for calculation of free space loss is based upon a point to point link model. This is because of the highly directional nature of MMDS line of sight propagation. It is quite a similar model to that of the microwave link system used in broadcast distribution systems and telecommunication networks.

The free space loss of a point to point radio frequency link can be given by the following expression:

$$Free\ space\ loss(dB) = 10\log_{10}\left(\frac{4\pi d}{\lambda}\right)^2 \qquad (12.1)$$

Where d is the path length and is expressed in meters, and λ is the wavelength of the electrical signal, also expressed in meters. Note that the path loss is heavily dependent on the path length between the transmitter and receiver. This equation can be rearranged in the following manner to derive the more common form of free space path loss equation,

$$Loss(dB) = 20\log_{10}\frac{4\pi d}{\lambda} = 20\log_{10}4\pi d + 20\log_{10}\frac{1}{\lambda}$$

Now

$$c = f\lambda \qquad (12.2)$$

where c, the speed of light, is a constant $[c \approx 3\times10^8]$, and f is the frequency of the electrical signal [Hz].

(12.1) can now be rearranged using (12.2) to substitute for λ.

$$\begin{aligned}
Loss(dB) &= 20\log_{10}4\pi d + 20\log_{10}\frac{f}{c} \\
&= 20\log_{10}4\pi + 20\log_{10}d + 20\log_{10}f + 20\log_{10}\frac{1}{c} \\
&= 21.9 + 20\log_{10}d + 20\log_{10}f + (-169.5) \\
&= -147.6 + 20\log_{10}d + 20\log_{10}f
\end{aligned}$$

Now if F is stated in units of megahertz $(f = F \times 10^6)$, and D is stated in units of kilometers $(d = D \times 10^3)$, then this equation can be rearranged into the more popular form,

$$Free\ space\ loss(dB) = 32.5 + 20\log_{10}D + 20\log_{10}F \qquad (12.3)$$

(12.3) is commonly used when calculating link budgets for microwave multipoint distribution systems, and for other microwave links. This free space model is based upon the concept of an expanding spherical wavefront as the signal radiates from a point source in space. It is used in satellite and deep-space communication systems where the signals truly travel in free space. In a terrestrial communication system additional losses, impairments, and obstacles, such as those described earlier in Table 12.1 and Section 12.2.1, may be expected.

12.4 MMDS Receive Installations

MMDS receive installations consist of a microwave antenna, coupled to a block down conversion unit (see Figure 12.1). This so-called block down conversion unit shifts the receive channel block from the 2.5 GHz band to another frequency block, within the UHF band for direct connection to the domestic television receiver. As the signal level is usually very low at any receive location, amplification of the signal is performed between the receive antenna and the block down converter. This external amplification is needed to raise the signal level prior to connection to a domestic television receiver. The down converter unit is a low noise device that increases the carrier-to-noise ratio and also increases the terminal voltage that is presented to the domestic receiver. The quality of the down conversion unit may be expressed in terms of a parameter known as the noise figure, which is the degradation in signal to noise ratio caused by passing a signal through the unit. It can be expressed as a ratio but is usually expressed in decibels (dB) (see Section 12.4.2).

12.4.1 Receive Equipment Specifications

As mentioned earlier, the receive equipment consists of a microwave antenna and a down conversion unit. The receive antenna has a typical gain of approximately 20 dB (relative to an isotropic radiator) and a half power beamwidth of 20°. It has a good front to back ratio of typically 25 dB.

The down conversion units can be either integrated into the antenna or be a completely separate unit. They are typically fabricated from gallium arsenide (GaAs), and have a typical gain of 30 dB, and a low noise figure of approximately 4 dB. The down conversion units are generally selected to produce an output at a particular UHF frequency block. Because active pre-amplification of the receive signal is required with MMDS, it is necessary to provide DC power to the down conversion unit. Adding a DC signal to the coaxial receive cable does this. The DC signal is derived from a domestic AC power point. As a result of this technical installation the cost to consumers is greater than with UHF and VHF terrestrial installations.

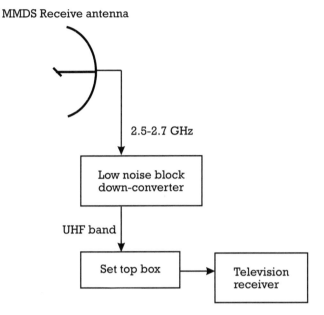

Figure 12.1   MMDS receive equipment.

12.4.2 Noise Figure

The noise figure (F_n) of a receiver (down converter or amplifier) is defined as the ratio of the signal to noise power supplied to the input terminals of the receiver to the signal to noise power at the output terminals. The calculation of noise figure is shown in (12.4) below.

$$F_n = \frac{s/n\_\text{input}}{s/n\_\text{output}} \qquad (12.4)$$

Where *s/n* is the signal to noise ratio.

It can be seen from (12.4) that the noise figure of an ideal down converter would be unity, that is it would introduce no noise of its own. However in real systems some noise will be added by the down converter and hence signal to noise ratio will be reduced slightly.

The noise figure is usually expressed in decibels and is simply:

$$NF = 10\log_{10} F_n \qquad (12.5)$$

Where *NF* is the logarithmic value of F_n.

12.4.3 Environmental Factors Affecting MMDS Receivers

Some important environmental factors affect receive installations for all transmission media, but in particular affect active electronic elements such as those found in satellite and MMDS receivers. Corrosion and sea salt spray affect outdoor hardware in coastal areas. All units will allow some moisture to penetrate as they heat and cool. Adequate sealant protection on active units and coaxial connectors must be used to prolong the life of the receive equipment. All down converter units should also employ remote regulation and surge protection in case of lightning strikes, and power line transient voltages.

12.4.4 Interference From Microwave Ovens

The microwave oven band is an internationally agreed standard and is based upon the resonance frequency of water dipoles. It is 2.450 GHz ± 50 MHz. Microwave oven units run at considerably high power levels (1 kW), and many have been found to leak significant amounts of microwave energy by RF receiver standards. They are by now ubiquitous in modern homes. Most MMDS networks operate just above this band with broadcasts between 2.5 and 2.7 GHz. However, MMDS down converter units have good rejection circuits and are generally not interfered with by these sources, unless there is a fault in the component, or the oven leakage increases.

12.5 MMDS Modulations

Many different types of modulation have been proposed for digital MMDS, including QPSK, 16-, 32-, and 64-QAM single carrier based systems. Multicarrier modulations have also been proposed for digital MMDS, including DVB-T compliant COFDM systems. In the United States the 8-level VSB system has also been proposed for transmission of digital signals. All of these modulations have been described in detail in Chapters 6 and 7.

All of these different modulations have been tested and shown to work adequately in MMDS networks [3, 4], generally because of the highly directional nature of the receive installations. This permits the receiver to minimize the level of multipath signal arriving at the receive antenna together with the main signal component. However, some field trials on digital MMDS have indicated that multicarrier systems have improved performance over single carrier systems in the presence of multipath signals, and in adjacent channel operation.

The DVB project has proposed standards for the transmission of digital signals on MMDS networks. The first of these is based upon the DVB-C standard, which is a 64 QAM single carrier modulation, and used in cable networks. It has been proposed for MMDS networks operating at frequencies below 10 GHz. It is known by the acronym DVB-MC [5].

The second DVB standard is based upon the DVB-S standard and uses the QPSK single carrier modulation proposed by the DVB project for satellite transmission of digital signals to the home. It has been proposed for MMDS networks operating at frequencies above 10 GHz. It is known by the acronym DVB-MS [6]. Another DVB standard has been developed by the DVB project for so-called multipoint video distribution systems (MVDS), which operate on a harmonized frequency band between 40.5–42.5 GHz. This system uses the same modulation as the DVB-MS system, but the bandwidth approximates that of the DVB-C system.

12.6 MMDS Transmitter and Combining Systems

12.6.1 Analog MMDS Arrangements

Analog MMDS transmitters normally produce two output signals, a visual carrier that includes a vision and a color subcarrier, and a frequency modulated sound or aural carrier. These two carriers must be efficiently combined, to minimize the degradation to either of these carriers, and to minimize the losses in the addition process. The combining of the visual and aural carriers is commonly referred to as diplexing, or CIND diplexing, where CIND refers to maintaining a constant impedance. The transmitter output is at the final channel frequency and power level for both the aural and visual components. As well as combining the aural and visual components, the diplexer also acts as a channel filter and is generally coupled to a wideband waveguide, which allows the addition of multiple channels. The typical transmitter powers range from 10 Watts to hundreds of Watts. Solid state amplification is generally used within the transmitters.

12.6.2 Digital MMDS Arrangements

Digital MMDS transmitters produce only one output signal which is normally output as the visual carrier, but which contains all of the information signals, in a digital format. This digital signal is at the final channel frequency and at the transmitter rated output power. The signal is generally passed through a channel filter that is coupled directly to a waveguide. It must be combined with other transmitter output signals before feeding the broadcast antenna. Solid state amplification is used generally, and the various transmitter architectures that are possible are similar to those described in Section 8.3.

12.7 Link Budgets

Link budgets are used to calculate the carrier-to-noise ratio (for digital systems) which is likely to be found at a particular receive location, if a signal is broadcast in a particular manner. They are very useful when trying to estimate the likelihood of high quality signal reception at a particular site. They require knowledge about the transmission system, the broadcast power and frequency, the path length, and the required-carrier to-noise ratio (C/N) for error free reception (see Section 9.4 for a treatment of BER versus C/N ratio).

12.7.1 ERP

The emitted radiated power (ERP) of a transmission system depends on the transmitter power, the antenna gain and the losses in the transmission system.

$$ERP = Tx\ power - Tx\ feeder\ losses + Tx\ antenna\ gain \qquad (12.6)$$

Where Tx is the transmitter, and the feeder is generally a low loss transmission line feeding the signal to the broadcast antenna.

Then the carrier to noise ratio can be derived from the following (12.7) [7]:

$$C\,/\,N = ERP - Path\ loss + Rx\ antenna\ gain - Rx\ feeder\ losses \\ -LNB\ noise\ figure + noise\ floor \qquad (12.7)$$

Where the path loss (FSL) is calculated from (12.3) above, Rx is the receiver LNB is the low noise block down-conversion unit described in Section 12.4.

12.7.2 Thermal Noise Floor and Noise Temperature of a Receiver

The thermal noise floor (*Ni*) of the receiver may be calculated using the following equation,

$$Ni = kTB \qquad (12.8)$$

Where B is the bandwidth of the broadcast channel, k is Boltzmann's constant = 1.38×10^{23} W/Hz/K (or -228.6 dBW/Hz/K), T is absolute temperature, measured in degrees Kelvin.

The concept of noise figure is not the preferred measure of noise for UHF and microwave low noise devices. A more convenient measure is the noise temperature [8], which is used extensively for low noise microwave amplifiers and antennas, and in this case it can be substituted into the above equation for the noise floor of the MMDS receiver, yielding:

$$Ni = k(T_{eq} + T_0)B \qquad (12.9)$$

Where T_0 is defined to equal 17 C = 290K and T_{eq} = equivalent noise temperature of a receiver with noise figure F. F in this case is a ratio, and not expressed in decibels.

Finally there is an equation which relates these parameters

$$T_{eq} = T_0(F - 1) \qquad (12.10)$$

From (12.9) and (12.10) it is possible to calculate the noise floor of the receiver and then to calculate the carrier-to-noise ratio from (12.7).

Example 12.1

A 1W digital transmitter feeds a signal to an MMDS antenna with 18dBi gain. The feeder losses are 5.8 dB. The broadcast frequency is 2.7 GHz. There is a receive antenna with 19 dB gain, located 30 km away, with 1 dB feeder cable losses and a low noise block down-converter with noise figure of 4 dB. The broadcast channel is 7.6 MHz wide. What is the resulting carrier to noise ratio?

To calculate the ERP, firstly convert figures to logarithmic units:

$$1 Watt = 10 \log_{10}(1) = 0 \, dBW$$

$$ERP = 0 - 5.8 + 18 = 12.2 \, dBW$$

$$FSL = 32.5 + 20\log_{10}(30) + 20\log_{10}(2700) = 32.5 + 29.5 + 68.6 = -130.6\,dB$$

$$Ni = 1.38 \times 10^{-23}(T_{eq} + 290) \times 7.6 \times 10^6$$

$$T_{eq} = T_0(F - 1) = 290\left(10^{\frac{NF}{10}} - 1\right) = 290(2.51 - 1) = 438\,Kelvin$$

Now, the noise floor can be calculated:

$$Ni = 1.38 \times 10^{-23}(438 + 290) \times 7.6 \times 10^6 = 7.63 \times 10^{-14}\,Watts = -131.1\,dBW$$

So the resulting carrier to noise ratio can finally be calculated:

$$C/N = 12.2 - 130.6 + 19 - 1 - 4 - 131.1 = 25.7\,dB$$

12.8 Implementation Scenarios for Digital MMDS

Digital MMDS is being implemented throughout America, Europe, Asia, and in parts of the Middle East. The challenge for network designers is to allow two systems, one the original analog system, and the new digital system, to operate in parallel within the same allocated frequency band. For example if the band from 2.5–2.7 GHz is used for analog MMDS already then it may be possible to interleave digital services between the existing analog channels.

In many of the systems deployed to date, only alternate channels are used for the existing analog services. There are approximately 22 channels, each being of 8 MHz bandwidth, that could be accommodated within the 2.5–2.7 GHz band. However, if only alternate channels are used for the present analog system, then there are 11 unassigned channels available at each MMDS cell [9]. Normally if odd/even channels are used for analog services then it is possible to interleave digital services between these channels using the even/odd channel numbers, as shown in Figure 12.2.

However in some countries, notably the United States, often all the channels are used for existing analog services, with two different transmission systems being used, one using the odd channel numbers, and a parallel system using the even numbered channels. If this is the case other alternatives must be considered, including removing some analog services and replacing them with more spectrally efficient digital services.

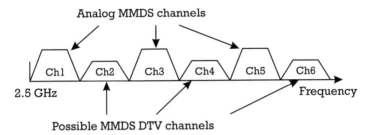

Figure 12.2 Possible implementation scenarios for digital MMDS services.

References

[1] Tesh, K., "Terrain Factors Part 2," *Wireless Broadcasting Magazine*, Vol. 2, Issue 9, November 1994, pp. 38–40.

[2] CCIR Rec. 525-1, "Calculation of Free-Space Attenuation," 1978–1982, pp 3–8.

[3] O'Leary, S., F. Kerrin, and J. Caffrey, "COFDM Trial on Wireless Cable System," *WCAI Technical Symposium*, Florida, February, 1995.

[4] O'Leary, S., and K. Cleary, "Transmission of Digital Signals Over MMDS," *IBC*, September, 1995.

[5] EBU/ETSI JTC, *Digital Broadcasting Systems for Television, Sound and Data Services; Framing Structure, Channel Coding and Modulation for Multipoint Video Distribution Systems at 10 GHz and Above*, EN 300 748.

[6] EBU/ETSI JTC, *Digital Broadcasting Systems for Television, Sound and Data Services; Framing Structure,Channel Coding and Modulation for Multipoint Video Distribution Systems Below 10 GHz*, EN 300 749.

[7] Hope, P. W., "Multichannel Microwave Distribution Service (MMDS) for Television Programme Distribution," *Cable Television Engineering*, Vol. 14, No. 2, March 1988, pp. 493–497.

[8] Kennedy, G., *Electronic Communication Systems*, McGraw Hill, 3rd ed., 1984.

[9] Caffrey, J., "MMDS (Wireless Cable): An Alternative Delivery Medium for Digital Terrestrial Television," *IBC*, 16–20 September, 1994.

Contents

Digital Audio Broadcasting and Digital Radio

13.1   Introduction

Currently there is much interest in digital audio broadcasting (DAB) and digital radio throughout Europe and America. There have been enormous advances made within the last few years in digital audio technologies, in particular digital recording, storage, and transmission systems. The development of DAB technology has been supported by the European Broadcasting Union (EBU), and in conjunction with the Eureka DAB project, EU-147, and the International Telecommunication Union (ITU), it has driven the development of DAB in Europe and elsewhere. Digital radio, and in particular DAB, is seen as a milestone in the development of digital broadcasting technology by broadcasters and manufacturers alike. The challenge to date has been to obtain public interest in this new technology. This chapter will discuss the reasons behind the interest in DAB technology by the broadcasting industry, the technical aspects of implementing a DAB system, and the issues involved in allocating bandwidth, error protection, and data casting capability to these digital services. It will also describe the state of the art proposals for other digital radio systems in the traditional analog radio broadcasting bands. It will describe the various groups that are working in the development of digital radio standards in the United States, Europe, and around the world.

13.2   DAB Organizations

There are many manufacturers committed to the development of DAB products, including transmitters and receivers, encoders and professional equipment, as well as PC card receivers and car radios. Many European service providers, network operators, and broadcasters are experimenting with pilot networks or have even established networks. In Europe, the United States, and elsewhere a major focus of DAB activity is through the activities of the WorldDAB forum. This organization is now regarded as a central point for the dissemination of information, news, and views on issues related to DAB. It is an organization that also manages conferences and symposia on digital audio broadcasting with the aim of promoting the development of this new technology. Also in the United States another organization, the USA digital radio group is working toward the development of a new DAB based digital radio technique (see Section 13.13).

13.3   Why Go Digital for Radio?

Analog radio since the early 1950s has been using the frequency modulated (FM) stereo system that provides high quality radio services to the listener. This system was designed originally for a static receiver connected to a moderate gain rooftop aerial which was polarized (usually horizontal). However, with technological and social developments the nature of radio listening has changed dramatically from the fixed receiver installation, to the now commonplace portable type receiver, and also the car radio receiver. Both of these radios utilize whip-type omnidirectional aerials that offer no protection against multipath signal reception. The existing FM radio networks are, of course, able to serve most of these portable and mobile listeners (especially if the signals are vertically polarized), but many countries have reported areas where reception is impaired by multipath interference and selective fading [1]. Amplitude modulated (AM) radio in the medium wave and long wave bands is well known for severe fading and interference problems, and as a result much development work is currently under way to use robust digital modulations in parallel with existing AM broadcasts. It is hoped that these digital modulations will improve the quality of the service that is received and eliminate most of the annoying interference effects. As we will see later in Section 13.12 it is intended to use existing high power analog transmitters for simultaneous analog (AM) and digital broadcasts.

Increasingly radio broadcasters wish to add value to the signals that they are broadcasting to the public. This can be achieved in many ways with digital radio, including the use of audio content related datacasting. Datacasting is a popular technique, where for instance, advertising images can be carried within the multiplex that carries the audio signals. With the emerging large liquid crystal display screens of digital radios it is also possible to display pictures, or text, that may be related to the audio content. Digital radio provides the broadcaster with a new tool to associate data and program material in an effort to increase the depth and impact of the audio content. It also can be used for commercial applications, including advertising as mentioned earlier, or to provide text-based services such as share prices or weather reports, to targeted audiences. Digital radio offers the broadcaster the opportunity to embed hypertext mark-up language (HTML) pages, as found on the Internet, and possibly JPEG encoded digital pictures into the multiplex. Data in this example is nonaudio information, and it can be related to the audio program stream, and termed program associated data (PAD). The data can also be completely independent of the audio program that is contained in the same multiplex, and in this example the stand alone services are adjustable to the capacity requirements of the channel. These multimedia objects are packetized using a standard known as the multimedia object transfer standard (MOT) prior to multiplexing onto the DAB multiplex. Many broadcasters are beginning to use the MOT standard to develop multimedia applications for mobile reception via DAB receivers [2].

Most broadcasting organizations have some degree of interest at present in DAB, ranging from only a limited interest, to full commitment to the deployment of the technology. There are many reasons for adopting digital audio broadcasting, and some of these reasons include, the opportunity to provide so-called CD quality audio for the first time to the radio listener, or the potential to eliminate multipath fading problems related to poorly covered areas by establishing single frequency networks. A single frequency network can be established in a similar way to that discussed earlier in Chapter 10, and this is because of the use of spectrum efficient technologies with DAB such as COFDM. As a result of using digital MPEG-based compression technologies when encoding the analog audio signals, it is possible to carry more than one service on any broadcast channel and this provides the broadcaster with a potential for new service provision on the same assigned frequency. It has been found in some countries that have adopted DAB, that typically better performance in mobile environments has been noticed than that of traditional VHF/FM radio services. As the number of car radios increases and the number of hours spent by people travelling in these cars also increases, this market is becoming quite significant to broadcasters.

13.4   Eureka 147 Project

The Eureka 147 project was a European initiative established in conjunction with the European Broadcasting Union (EBU) to devise a digital audio broadcasting system which would be able to deliver high quality audio signals to even mobile receivers in the presence of multipath propagation conditions. It was hoped that a worldwide standard for digital audio broadcasting would emerge from the work of this project. This group developed the COFDM technology, which has been the subject of discussion in earlier chapters dealing with terrestrial digital television, in response to this need. Digital circuits have also been developed by the project group to allow data compression based on the psycho acoustical characteristics of the human ear to remove redundancy in the audio information signal. Techniques were developed to multiplex different services together onto the same data stream, including data signals and pictures. Error correction algorithms were developed to make the system more robust. Extensive field testing in different European countries in conjunction with the EBU led to the CCIR adopting draft recommendations in 1991 from the project group on the requirements for a digital audio broadcasting system [3].

13.5   The DAB System

The DAB system comprises a number of elements that are related and which are needed to implement a complete digital radio transmission chain. These include analog audio signal sampling, digital compression, data encoding, digital signal multiplexing, time stamping of the ensemble signal, digital distribution via telecommunication circuits, channel encoding and modulation, transmission, and finally reception. Some of the more important aspects of the DAB system will be dealt with in the subsequent sections. Particular emphasis will be placed on the terrestrial DAB system, referred to as T-DAB. Figure 13.1 below details some of the more important aspects of a terrestrial DAB network. It illustrates the concept of a DAB service provider who provides a signal for inclusion in the DAB multiplex, which may contain audio, and a data signal. Also shown is the ensemble multiplexer, which gathers together services from possibly many service providers and transports the ensemble to transmitter sites over typically a digital telecommunications network. The DAB signal is then channel coded and broadcast to the consumer. A comprehensive European Telecommunication Standard (ETS) has been developed which details a complete DAB transmission system, and in particular contains refinements to the Eureka 147 DAB system with regard to establishing single frequency networks in the L-band (1.4 GHz), sampling at low bit rates, establishing conditional access systems, and other service information related matters [4].

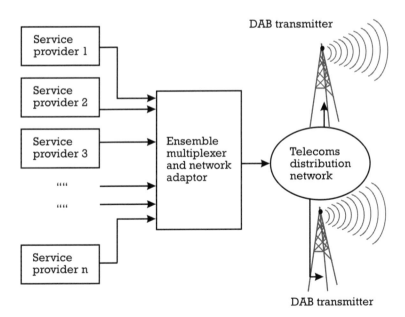

Figure 13.1   DAB network schematic.

13.5.1 DAB Transmission Modes

Four different DAB transmission modes have been defined within the European standard, mentioned above, to allow operation of the system within different radio frequency bands, with different network configurations, and with different transmission parameters. The details of the different transmission modes are shown below in Table 13.1. The overall bandwidth of the system is approximately 1.536 MHz irrespective of the mode of operation chosen. Also the system uses differentially encoded quadrature phase shift keying (D-QPSK) as the modulation in conjunction with COFDM channel encoding to form the radio frequency broadcast signal. For a detailed description of QPSK modulation the reader is referred to Sections 6.4 and 13.9, and for more information about COFDM technology refer to Chapters 7 and 10.

13.6 European Frequency Allotments for DAB

In July 1995 a planning meeting for terrestrial digital audio broadcasting (T-DAB) was held under the auspices of CEPT (European Conference of Postal and Telecommunications Administrations) in Wiesbaden to agrede a frequency allotment plan and associated procedures to facilitate the introduction of T-DAB. It was attended by delegates from 35 CEPT

Table 13.1
DAB Transmission Modes

DAB Mode	No. Of carriers	Carrier spacing (kHz)	Transmission frame duration (ms)	Total Symbol duration (ms)	Guard interval (ms)	Intended frequency bands	Type of service
I	1,536	1	96	1,246	246	I, II, III	Terrestrial SFN
II	384	4	24	312	62	{I, II, III, IV, V, L-band}	Local services
III	192	8	24	156	31	Below 3 GHz, and cable	
V	768	2	48	623	123	{I, II, III, IV, V, L-band}	Local services and L-band SFN

administrations, representing member countries of the International Telecommunication Union (ITU), and including all European Union (EU) member countries and the Russian Federation.

At this meeting frequency allotments were agreed on for the different countries to allow introduction of T-DAB in different bands while minimizing interference to other radio frequency users. A procedure was also agreed to allow the administrations to convert these block frequency allotments into frequency assignments at particular sites. It is generally accepted that this meeting was a major milestone in the development of T-DAB and the agreement is referred to as the Wiesbaden agreement [5]. The agreement allocated channels for T-DAB within different frequency bands, including band I (47– 68 MHz), band II (87.5–108 MHz), band III (174–240 MHz), and the L-band (1452–1492 MHz).

Within the agreement a harmonized channelling plan was formed, where each T-DAB signal was arranged into frequency blocks for a number of bands. At a particular frequency within a selected band a T-DAB block was given a designation. Each designated block has a reference number and has a common bandwidth of 1.536 MHz. There are guard bands on either side of each T-DAB block to minimize interference to other adjacent T-DAB services and these are typically 176 kHz wide. Some guard bands have a larger bandwidth, up to for example 336 kHz width at the edge of a band where extra protection to other services is needed.

Some of the T-DAB frequency blocks are shown in the following Tables, 13.2, 13.3, and 13.4. The more commonly used frequencies are found both in the L-band frequency blocks, and also in the VHF (band III) frequency blocks.

Note that in all of these tables, there are 4 frequency blocks associated with each (7 MHz wide) channel, for example in channel 5 the blocks 5A, 5B, 5C, and 5D are contained within a standard 7 MHz television channel (174–181 MHz). Guard bands are typically employed on either side of the T-DAB channel with a typical bandwidth of 176 kHz. However there are 6 frequency blocks associated with channel 13 above, which in this case is 10 MHz wide and has narrow guard bands (32 kHz wide) between blocks 13 C and 13 D (see Table 13.3).

The frequency blocks contained in the tables are all 1.536 MHz wide. The most common implementations are within the VHF band III, which is used to implement terrestrial single frequency networks, and the L-band, which is used for localized networks and regionalized single frequency networks. A typical channel arrangement is shown in Figure

Table 13.2
T-DAB Frequency Blocks for VHF Band I

T-DAB block number	Center frequency (MHz)	Bandwidth (MHz)
2A	47.936	1.536
2B	49.648	1.536
2C	51.36	1.536
2D	53.072	1.536
3A	54.928	1.536
3B	56.64	1.536
3C	58.352	1.536
3D	60.064	1.536
4A	61.936	1.536
4B	63.648	1.536
4C	65.36	1.536
4D	67.072	1.536

13.2 for T-DAB channel 12, illustrating how a 7 MHz television channel (a common European VHF television channel bandwidth) is divided into 4 individual T-DAB frequency blocks for DAB broadcast channels, and how it provides guard bands between each T-DAB channel.

13.7 DAB Signal Structure

The DAB system allows the simultaneous carrying of many audio services together with data services, including Internet pages and digital pictures. In mode I of DAB, which is used for terrestrial broadcasting, the DAB signal consists of frames, each of which has a 96 ms duration and contains a total of 77 symbols. There are three channels associated with the DAB signal multiplex and these include the main service channel, the Fast information channel, and the synchronization channel. A general DAB signal structure is shown below in Figure 13.3 and is independent of the DAB mode selected.

There is also a null symbol that is broadcast at the start of every DAB frame. A null symbol is the first OFDM symbol of any transmission frame, and can vary in length from 168–1,297 μsec, depending on the transmission mode chosen. During the null symbol, the DAB transmitter is

Table 13.3
T-DAB Frequency Blocks for VHF Band III

T-DAB block number	Center frequency (MHz)	Bandwidth (MHz)	T-DAB block number	Center frequency (MHz)	Bandwidth (MHz)
5A	174.928	1.536	10A	209.936	1.536
5B	176.64	1.536	10B	211.648	1.536
5C	178.352	1.536	10C	213.36	1.536
5D	180.064	1.536	10D	215.072	1.536
6A	181.936	1.536	11A	216.928	1.536
6B	183.648	1.536	11B	218.64	1.536
6C	185.36	1.536	11C	220.352	1.536
6D	187.072	1.536	11D	222.064	1.536
7A	188.928	1.536	12A	223.936	1.536
7B	190.64	1.536	12B	225.648	1.536
7C	192.352	1.536	12C	227.36	1.536
7D	194.064	1.536	12D	229.072	1.536
8A	195.936	1.536	13A	230.784	1.536
8B	197.648	1.536	13B	232.496	1.536
8C	199.36	1.536	13C	234.208	1.536
8D	201.072	1.536	13D	235.776	1.536
9A	202.928	1.536	13E	237.488	1.536
9B	204.64	1.536	13F	239.2	1.536
9C	205.352	1.536			
9D	208.064	1.536			

effectively off, with no signal output. As a result it has become popular to use the null symbol as an opportunity to broadcast transmitter identification information (TII). This information may be broadcast using a few fixed and known carriers from the DAB ensemble. It is used in SFNs to identify transmitters and give some measure of the channel response (see Section 13.7.3).

13.7.1 The Main Service Channel, Common Interleaved Frames, and Capacity Units

The output signal from the main service multiplexer is contained in the main service channel (MSC) part of the transmission frame. It is a serial digital signal and is common to all four DAB transmission modes. The

Table 13.4
T-DAB Frequency Blocks for L-Band

T-DAB block number	Center frequency (MHz)	Bandwidth (MHz)
LA	1452.96	1.536
LB	1454.672	1.536
LC	1456.384	1.536
LD	1458.096	1.536
LE	1459.808	1.536
LF	1461.52	1.536
LG	1463.232	1.536
LH	1464.944	1.536
LI	1466.656	1.536

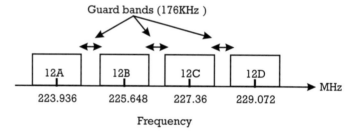

Figure 13.2   Channel 12 T-DAB frequency blocks.

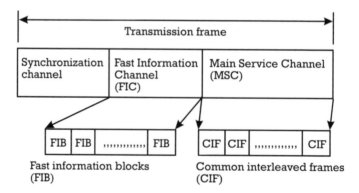

Figure 13.3   Description of the DAB signal structure.

main service channel is made up of a sequence of common interleaved frames (CIF). Each CIF contains 55,296 bits, which is equivalent to 864 capacity units (CU), where each so-called capacity unit contains 64 bits of information and is the smallest addressable unit within the DAB signal. The main service channel contains 72 symbols of the total 77 symbol frame.

Integral numbers of these capacity units can be grouped together to form subchannels of the main service channel. Each subchannel can, for example, carry a different audio service. Each subchannel is made up of an integral number of capacity units and is individually error protected using a different rate convolutional code. This means that different services within the same multiplex can be error protected to different levels (see Section 13.9.3). Each service within a DAB multiplex can have therefore a different number of capacity units assigned to it. However the sum of all of these sub-channels together with the overheads introduced in error protecting the services must not exceed the maximum data capacity of 864 capacity units mentioned above.

13.7.2 The Fast Information Channel (FIC)

This channel is part of a transmission frame and contains information about the configuration of the actual multiplex, the data service components, and any service information, if present. It is common to all transmission modes. It is made up of fast information blocks (FIB), each of which contains 256 bits of information about the configuration of the main service channel. Each information block is transmitted with a high level of error protection because it is used to reconfigure the multiplex if needed. The fast information channel is transmitted within three symbols of the total 77 symbol frame.

13.7.3 The Synchronization Channel and Null Symbol

The synchronization channel is used to reference the time information in the fast information channel. This information contains the time of transmission of the null symbol in the transmission frame carrying time information. As mentioned earlier, the null symbol is an interruption to the transmission lasting up to possibly 1.297 ms, which is broadcast in every frame and allows a receiver to lock onto the DAB signal. It is also used for demodulation functions such as channel state estimation, automatic frequency control, and may contain transmitter identification information.

13.8 Distributing DAB Signals on Telecommunication Networks

There is a European Telecommunications Standard (ETS) for the distribution of DAB signals from, for example, a studio to terrestrial main station broadcast sites, where channel coding and broadcasting may take place [6]. These distribution circuits are generally digital telecommunication networks, which use telephony standards and interfaces. As a result a standard was needed to allow the adaptation of a DAB ensemble, carrying subchannels of audio and other data, to a format compatible with switched telecommunication networks. The standard developed also allows for the correction of network transit delay when establishing single frequency networks, as discussed in Chapter 10 earlier.

13.8.1 Ensemble Transport Interface (ETI)

The ensemble transport interface (ETI) is a generic title for a DAB signal which contains the multiplexed ensemble of services prior to channel coding (COFDM encoding) at a transmission site. It can have different physical realizations depending on where it is found in the network (see Figure 13.4). At the simplest level, the signal is referred to as a logical interface (LI) and designated as ETI(LI), but it has no actual physical definition. It consists of frames, each of which contains approximately a 24 ms period of the DAB signal and consists of a status and a data field. The data field contains the information payload and it remains unchanged no matter what physical interface is used. The status field contains information about the quality of the network and its contents may be changed by different physical interfaces.

13.8.2 Network Independent (NI) Layer and ETI(NI, G.703)

The simplest physical interface for the ETI just described is as a network independent layer. In this form the DAB ensemble can be passed from equipment to equipment within the same local area. The interface is useful for interconnecting equipment within the same location on cables, and is designated as ETI(NI). It is not a suitable interface for DAB signal distribution on a telecommunications network. It can also be designated as ETI(NI, G.703) where G.703 refers to the local interface used in the network independent layer. ETI(NI, G.703) uses G.703 line coding, carrying the data and clock information on the same 2.048 Mbps serial connection [7]. No extra error protection is added to the ETI(LI) by this physical adaptation and it should not therefore be used for distribution

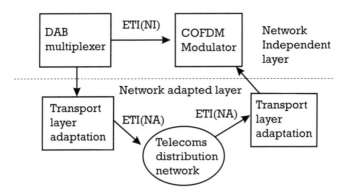

Figure 13.4   Ensemble transport interfaces ETI(NI) and ETI(NA).

on standard telecommunication networks. The author finds the usage of G.703 problematic, as many unaware users may assume that this physical interface may be suitable for use on a standard telecommunications 2.048 Mbps circuit. However, as well as not having adequate error protection capability, the ETI(NI, G.703) interface is not suitable for transmission over a network that might detect a frame padding field, or a long string of ones, as an alarm indication signal (AIS).

13.8.3   Network Adapted (NA) Layer and ETI(NA, G.704)

To allow the transport of a DAB ensemble signal over a non secure transmission network such as a telecommunications network, extra protection is needed above that provided with the NI layer, and as a result a different physical interface has been defined. This interface is known as a *network adapted* (NA) layer and is designated as ETI(NA, G.704). The designation G.704 refers to an ITU recommendation defining telecommunication framing structure [8]. The extra protection is provided before connecting the DAB signal to any telecommunications equipment. The interface allows the addition of time stamps (similar to the MIP described in Chapter 10) to allow for the different network transport delays to remote transmitter sites. This allows the network operator to synchronize all broadcasts of data, as is required in establishing a single frequency network. It also uses Reed-Solomon forward error correction coding to combat any errors generated while the DAB signal is in distribution. The physical characteristics of the ETI(NA, G.704) conform to the requirements of the ITU-T recommendation G.703. It is necessary to use a timing and a frequency reference signal at the output of the DAB ETI(NA,

G.704) interface equipment to generate this network adapted layer. This is an identical process to that required in establishing a DTV single frequency network.

13.8.4 Compensating for Transit Delays in Single Frequency Networks

Invariably there will be unequal delays within any terrestrial distribution network when transporting the ETI(NA) signals to remote transmitter sites for channel encoding and broadcasting. This is because the path lengths and distances between transmitters and the main distribution point will not be equal in a real network. If a single frequency network is to be established for DAB or for DTV, it is necessary to ensure that all broadcasts are bit identical and synchronized in time and frequency (see Chapter 10). The ETI(NA, G.704) frame structure has features that allow a single frequency network to be easily implemented.

At the multiplexer site, a global positioning system (GPS) can be used to add a frequency and time reference to the DAB signal. The timing reference used is a 1 pulse per second (PPS) signal and the frequency reference is typically a 10 MHz sine wave signal. Both may be readily obtained from a GPS receiver. A time stamp, which contains the local universal co-ordinated time (UTC) taken from the GPS receiver, and a variable timing delay, is added to the DAB signal at the multiplexer. The DAB multiplex manager allows the addition of a variable timing delay to the time stamp, to compensate for the maximum possible transit delay within the distribution network. In this way it is possible to ensure that all transmitters in the network wait until this delay has passed before broadcasting that particular frame of data. It requires the use of GPS receivers at the transmitter sites to synchronize the time delays.

13.9 DAB Modulation—DPQSK

DAB uses a very simple digital modulation technique known as differential quadrative phase shift keying (DQPSK or D-QPSK) which is a refinement of the QPSK modulation discussed in Section 6.4. In this variant of QPSK, differential encoding is applied to the data to form QPSK symbols on each carrier, and this results in a slightly different modulation, known as $\pi/4$-shifted D-QPSK modulation, being subjected to each carrier. This variant only differs from standard QPSK modulation in a minor way, which is that the phase of the reference signals are increased by 45° each symbol period [9].

With differential modulation the information is not carried by the absolute state of the carriers but in the transitions between states. The data is encoded in the magnitude and direction of the phase shift, not in the absolute position on the constellation. In some forms of differential modulation there are restrictions on the allowable transitions. By restricting the trajectory of the transitions, for example, to avoid passing through the origin, it is possible to simplify transmitter design.

In the same manner as QPSK it allows for 2 bits of information to be mapped onto each symbol, as shown in Section 6.4. A symbol can be thought of as a period of time where the phase of a carrier is kept constant. There are 1,536 separate carriers defined within DAB mode I (see Table 13.1). By using differential demodulation, it is easy to receive and decode a DAB signal, even when the modulation is removed every null symbol.

13.9.1 DAB Use of COFDM

As mentioned earlier, COFDM was developed within the Eureka 147 project for use as a digital radio channel coding system to combat frequency selective fading, and multipath propagation distortions. It also can be used to establish single frequency networks, and as a result it is attractive for digital radio, digital television, and in particular DAB.

Many of the COFDM parameters as discussed in Chapter 7 and Chapter 10 are relevant to DAB. The guard interval, the useful interval, and the total symbol duration are related in a similar manner as was shown for a COFDM DTV signal in (7.1). In transmission mode I of DAB, as shown in Table 13.1, the total symbol duration is approximately $1,246\,\mu s$. 246 ms of this symbol duration is used in the guard interval to combat multipath, leaving approximately 1 ms for data transmission in the useful symbol.

13.9.2 DAB System Capacity

As was seen in Table 13.1 earlier, the total symbol duration for a mode I, DAB signal is 1.246 ms (1,246 μs). The total symbol rate is simply the inverse of this figure,

$$\frac{1}{1.246 \times 10^{-3}} \approx 800\,symbols\,per\,second \tag{13.1}$$

The DAB signal is transmitted in groups of symbols called frames, and in this mode each frame is approximately 96 ms long. There are 77 symbols contained in each 96 ms frame and of these 72 symbols contain data payload. The remaining 5 symbols are used in the fast information channel and for synchronization of receivers. As was mentioned earlier each symbol can carry 2 bits of information with D-QPSK modulation, which means that each DAB frame of 96 ms duration contains (2 × 72) 144 bits of information, and the total data rate per carrier is therefore,

$$\frac{144}{96 \times 10^{-3}} = 1500\,bits\,/\,carrier \qquad (13.2)$$

$$The\,total\,data\,rate = 1500 \times 1536(carriers) \approx 2.304\,Mbits\,/\,second \quad (13.3)$$

The system capacity of the main service channel of a DAB multiplex is equal to the total data rate of approximately 2.304 Mbits/second. This is the maximum possible payload for a DAB multiplex, without the use of error correction.

The total data capacity of a DAB ensemble can be calculated in a similar manner as above and it includes the fast information channel and synchronization bits. It is calculated by using all 77 symbols in the above calculation instead of using only the 72 payload symbols. The maximum data capacity is therefore

$$\frac{77 \times 2 \times 1536}{96 \times 10^{-3}} \approx 2.4\,Mbits\,/\,s \qquad (13.4)$$

So the overall data carrying capacity of a DAB signal is approximately 2.4 Mbps, while the data payload (main service channel) of the DAB signal is approximately 2.3 Mbps, leaving just under 0.1 Mbps for the fast information channel and synchronization purposes. As we will see in later sections the DAB system has specified only a 2.048 Mbps telecommunications circuit for distributing DAB signals from studio to transmitter sites. This circuit would not appear to have the capacity to carry a 2.4 Mbps DAB signal, but as we shall see the DAB system has a method to reduce the bit rate of distribution signals down to rates which are compatible with this type of circuit.

13.9.3 DAB Error Protection
In DAB systems forward error correction (FEC) codes are used in a similar manner to those used in DTV, however in this case the error correction coding is applied to each service within a multiplex and not to the whole multiplex together, as is the case in DTV. This means that individual audio services within the same multiplex can have strong or weak error protection levels of coding applied to them. This process is

performed only after the multiplex has been distributed to the channel encoder (COFDM modulator) at the transmitter station. By not adding error protection to the DAB signal at the multiplexer, it is possible to maintain the gross bit rate of the DAB signal at a suitably modest bit rate. However the level of error protection must be selected at the multiplexer, and this sets the FEC rate used for each audio subchannel. The FEC rate is the approximate ratio of the information signal to the total data rate. For example a 256 Kpbs digital audio signal would generate approximately a 753 Kpbs data rate signal using a FEC rate equal to 0.34 (see Tables 13.5 and 13.6). This FEC rate information is sent within the fast information channel to the modulator to allow the addition of FEC at the modulator immediately prior to transmission. By using this approach it is possible to keep the data rate of any DAB signal below the standard telecommunications hierarchical bit rate of 2.048 Mbps. This means that a standard telecommunications circuit can be used to carry a DAB signal over a distribution network even though the maximum gross bit rate of a DAB signal (2.4 Mbps) exceeds that of a 2.048 Mbps telecommunications circuit. As we will see later adaptation is needed for a DAB signal to be carried on such a distribution circuit.

The actual error protection used in DAB transmission is a Viterbi error correction technique, which is a well-known convolutional coding scheme. Different levels of error protection are possible, and the DAB multiplex manager can assign protection levels to services as required. The overall data payload of the multiplex (including error protection overheads) cannot exceed however 2.304 Mbps. The protection levels are numbered from 1 to 5 as shown in Table 13.5, along with the corresponding FEC rates for a few different rate encoded audio signals. These audio encoding rates are among the more popular audio encoding rates used for DAB services today, where 256 Kbps is considered high quality stereo, and 96 Kbps is acceptable for a mono music service or speech using MPEG layer II audio encoding technology.

These protection levels can add a huge amount of overhead to a DAB service and increase the required bit rate to transmit the signal. Some illustrative calculations have been made, and the figures indicate that, for example, a 192 Kbps encoded audio signal with strong error correction, corresponding to level 1 protection would require a bit rate of approximately 565 Kbps and this would reduce the available payload for other services. Some approximate calculations are shown below in Table 13.6 for audio signals encoded at different bit rates and using the defined protection levels. It indicates the expected gross bit rate for a service within

Table 13.5
DAB Protection Levels for Various Audio Signals and Corresponding FEC Rates

Protection level	Audio signal encoded at 256 Kbps	Audio signal encoded at 192 Kbps	Audio signal encoded at 160 Kbps	Audio signal encoded at 96 Kbps
1	0.34	0.35	0.36	0.35
2	0.41	0.43	0.43	0.43
3	0.5	0.51	0.52	0.51
4	0.57	0.62	0.58	0.62
5	0.75	0.75	0.75	0.75

Table 13.6
Gross Bit Rates (Kbps) for Various Bit Rate Audio Signals, With Different Protection Levels

Protection level	Audio signal encoded at 256 (Kbps)	Audio signal encoded at 192 (Kbps)	Audio signal encoded at 160 (Kbps)	Audio signal encoded at 96 (Kbps)
1	753	565	444	274
2	624	447	372	223
3	512	376	308	188
4	449	310	276	155
5	341	256	213	128

the multiplex based on the protection level applied to that service. For example, to afford a 256 Kbps encoded signal with a high level of protection (level 1), would require a gross bit rate of 753 Kbps. This constitutes approximately 32% of the overall multiplex capacity, for just a single service.

13.10 MPEG Audio Layer

The audio subgroup of the Moving Pictures Expert Group (MPEG) has developed standards for the generic coding of pulse code modulated (PCM) audio samples with set sampling rates. There are two standards, MPEG-1 audio [10] and MPEG-2 audio [11], and they are very closely related. There are three different layers for these standards which are defined for different applications. (See Section 4.13 for more details on MPEG audio coding and MPEG layers.) In Table 13.7 a comparison of the main features of the MPEG-1, and MPEG-2 standards is given. Layer II was chosen for broadcasting applications including DTV and DAB.

From the above table it can be seen that the MPEG-2 audio coding scheme has enhanced the capability of MPEG to provide multichannel sound, including surround sound and multichannel audio broadcasting. It also has increased the capacity of the main service channel of DAB systems by using lower sampling rates on speech channels such as news services, while providing similar quality as that of MPEG-1 for these speech services.

13.10.1 Sampling Rate and Resolution of Encoder

In any audio encoder the most important features are the sampling rate and the resolution of the analog to digital converter. The number of bits that are used when setting the resolution of the analog to digital converter is a very important parameter in setting the quality of the encoder. For instance an 18-bit resolution analog to digital converter used in conjunction with a 32 kHz sampling rate would result in an uncompressed data rate of,

$$18 \times 32 \times 10^3 = 576 \, kbits \, / \, s$$

This is the bit rate for a mono service and it would be twice this rate for a stereo service. This bit rate is far too high for broadcast use, so compression is used to reduce this data rate down to a more typical DAB data rate of 192 Kpbs. This data rate reduction system uses principles of psycho-acoustics to remove irrelevant data from the signal. The data would not be heard by a human ear, due to the masking of sounds within

Table 13.7
Comparison of the MPEG-1 and MPEG-2 Audio Characteristics

Layer II	Maximum number of channels	Minimum sampling rate	Minimum bit rate	Maximum bit rate
MPEG-1 audio	2	32 kHz	32 Kbps	384 Kbps
MPEG-2 audio	5	16 kHz	8 Kbps	1066 Kbps

the signal by louder sounds at nearby frequencies. The topic of psychoacoustics was mentioned in Section 4.13.

13.11   Multiplexing Different Format Signals

A DAB multiplex can contain various formats of signals including analog audio, digital audio (AES/EBU), data, and even compressed video or still pictures. Each particular component service of the total ensemble can contain these signals. This is because each service component multiplexer can combine these signals, as shown below in Figure 13.5. The signal can then be independently passed onto another location for multiplexing onto the DAB multiplex. This means that a service provider can generate an independent service locally, and then transport the service to the ensemble multiplexer where the service can be combined with other services to form the final DAB ensemble.

13.11.1   Program Associated Data (PAD)

PAD is an independent data stream that is combined into the audio stream of a particular audio service. It is intended to complement the audio content of the service and contain data information that is relevant to the audio content. It is not subjected to the variable delay that other information channels can be subjected to, and it is synchronized to the audio data throughout the transmission chain. PAD information is under the control of the service provider and has a bit rate of typically 0.67 Kbps at 48 kHz sampling rate.

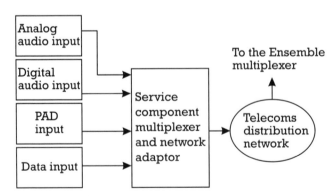

Figure 13.5   DAB service multiplexing prior to ensemble multiplexing.

13.12   Digital Radio in the AM Bands

There is currently much interest in broadcasting digital radio signals in the traditional amplitude modulation (AM) bands. This is because it is accepted that digital modulation would improve the quality of broadcasts in the long wave (LW), the medium wave (MW), and the short wave (SW) bands. Problems associated with broadcasting in these bands could be eliminated, including fading and interference. Audio quality would improve for the listener, and new types of services could be developed. Lower power transmitters could be used with very large coverage areas, so there would be cost benefits to the broadcaster. In America there is an organization called the USA digital radio group (USADR) which is developing digital transmission standards in the AM bands and in the FM band using the IBOC principles, as will be discussed in Section 13.13 below.

13.12.1   Digital Radio Mondiale ("Digital Radio World-Wide")

In 1998 a consortium was formed which now has over 30 members from different parts of the world, and is called digital radio mondiale (DRM). It is a sector member of the ITU with international study groups leading the development of a new standard, which is expected early in the 2nd millennium. It is studying two different candidate modulation schemes, which involve amplitude phase shift keying (APSK), and COFDM using QPSK to modulate the 184 carriers. The main features of both schemes are given in Table 13.8 below. Note that MPEG-2 is the chosen audio-encoding standard, and that layer II is that specified for this new technology [12]. Layer II or MUSICAM is that used presently for DAB.

13.13   In Band on Channel DAB (IBOC DAB)

In the United States a variant of DAB is being developed and has been licensed in some areas for broadcast trials. It is known as "in band on

Table 13.8
Features of the DRM Candidate Modulation Schemes for AM Radio

Modulation type	RF bandwidth	Audio bandwidth	Audio coding	Usable data rate
APSK	9 kHz	>8 kHz	MPEG-2, layer II	21 Kbps
COFDM (QPSK)	9 kHz	>8 kHz	MPEG-2, layer II	24 Kbps

channel DAB" (IBOC DAB) because it shares the same frequency bands
as frequency modulated (FM) signals. In this hybrid system the DAB signal is placed on either side of the FM radio signal, from about 130 kHz to
200 kHz away from the host FM center frequency. The total DAB power
in each sideband is set to typically −25 dB relative to the FM radio signal.
The IBOC DAB signal is broadcast from the same transmitter as is used for
the analog FM service. This makes it attractive to broadcasters who can
re-use existing equipment for analog and digital radio. The digital audio
information is transmitted in each sideband and only one sideband needs
be received to decode the signal. This signal redundancy not only protects
against corruption of one sideband but also increases the performance of
the system if both sidebands are decoded together. The COFDM carriers
are individually modulated using QPSK modulation and there are
approximately 95 carriers in each sideband [13].

The technique uses power multiplexing, which exploits the capture
effect of FM demodulation to separate signals at different power levels.
The capture effect is a well known effect in the FM demodulation of two
different level signals at nearby frequencies, which allows the demodulation of the stronger FM signal without noticeable interference from the
lower power one. In a FM/DAB configuration it is possible using this
principle to demodulate the strong FM signal, remodulate it, and subtract
it from the incoming signal. The resulting signal would then contain only
the lower power DAB signal [14]. This means that a DAB signal could be
power multiplexed over an existing FM radio station signal. Of course if
there is no DAB signal present the receiver would just demodulate
the analog signal. The U.S. digital radio group has developed the IBOC
system. However, it is still not yet known if the system will be adopted for
commercial use by any broadcasters.

The system is being tested at approximately 10 FM transmitter sites in
the United States until the beginning of the year 2000, when it is planned
to begin to commercialize the system. It differs from the earlier described
European DAB system in the areas of error correction, interleaving, and
audio coding. These changes were made to allow compatibility of the
IBOC DAB system with the existing analog FM and AM services.

References

[1] Lau, A., and W. F. Williams, "Service Planning for Terrestrial Digital Audio Broadcasting," *EBU Technical Review*, Summer, 1992, pp. 4–26.

[2] Schnaithmann, M., "Radio for Your Eyes, Datacasting Via DAB," *World Broadcast News*, June 1999, pp 54–56.

[3] O'Leary, T., "Terrestrial Digital Audio Broadcasting in Europe," *EBU Technical Review*, Spring, 1993, pp. 19–26.

[4] ETS 300 401, "Radio Broadcasting Systems; Digital Audio Broadcasting (DAB) to mobile, portable and fixed receivers," 1997.

[5] Final acts of the CEPT T-DAB Planning Meeting, Wiesbaden, 1995.

[6] ETS 300 799, "Digital Audio Broadcasting (DAB); Distribution Interfaces; Ensemble Transport Interface (ETI)," 1997.

[7] ITU-T Recommendation G.703, "Physical/Electrical characteristics of hierarchical digital interfaces: Section 6. Interface at 2.048 kbit/s," 1972.

[8] ITU-T Recommendation G.704, "Synchronous frame structures used at primary and secondary hierarchical levels: Section 2.3 Basic frame structure at 2.048 kbit/s," 1988.

[9] Shelswell, P., "The COFDM Modulation System: The Heart of Digital Audio Broadcasting," *Electronics & Communication Engineering Journal*, June, 1995, pp. 127–136.

[10] ISO/IEC 11172-3, "Coding of Moving Pictures and Associated Audio for Digital Storage Media at up to 1.5 Mbit/s —Audio Part," March, 1993.

[11] ISO/IEC 13818-3, "Information Technology: Generic Coding of Moving Pictures and Associated Audio—Audio Part," 1994.

[12] Senger, P., "Digital Radio in the AM Bands, Challenge or Chance for International Broadcasters," *IBC*, September 1998, pp. 614–618.

[13] Kroeger, B., and D. Cammarata, "Robust Modem and Coding Techniques for FM Hybrid IBOC DAB," *IEEE Trans. On Broadcasting*, Vol. 43, No. 4, December, 1997, pp. 412–419.

[14] Scalart, P., et al., "Performance Analysis of a COFDM/FM in Band Digital Audio Broadcasting System," *IEEE Trans. On Broadcasting*, Vol. 43, No. 2, June, 1997, pp. 191–198.

CHAPTER

14

Contents

Portable and Mobile Digital Broadcasting

14.1 Introduction

For some time now there has been much interest among broadcasters in a fast and reliable means of getting video and audio material from remote sites back to television studios. It is termed contribution broadcasting, or a contribution circuit, as the content is usually a contribution to the main program material. The growth of electronic news gathering (ENG) over the past few years for portable and mobile broadcasting, and in particular the emergence of satellite news gathering (SNG) for portable broadcasting has been driven by a need for cost effective, responsive, and universal outside broadcasting (OB) methods. The main applications to date have been news reporting and sports events, but more and more the traditional work of large outside broadcasting units (OBU) is been done by smaller SNG and ENG units, using either microwave links or satellite links.

Traditionally broadcasters have used analog microwave technology for ENG feeds from trucks and helicopters, for the provision of news and sports contributions. However these analog links suffer from interference and multipath problems that can degrade the quality of the contribution link. As a result there is interest in technologies that may reduce the susceptibility of these links to degradations.

There is also growth in the usage of digital technologies within SNG which is termed digital satellite news gathering (DSNG), and also there is beginning to be a usage of COFDM terrestrial based technologies for low cost mobile broadcasting applications. The main reasons for the above transitions to digital based technologies are that costs are now competitive, equipment is smaller and lighter, and equipment is reliable and rugged. There is also more compatibility between the digital acquisition equipment used in the field, and digital studio and digital transmission equipment.

As well as detailing the above-mentioned technologies for contribution broadcasting, this chapter will discuss important aspects when building vehicles for these applications. It will also refer to common climatic causes of ENG system failure and some preventative measures. It should be noted that some of the climatic and propagation effects described in the earlier chapter on digital MMDS are applicable to microwave ENG systems.

14.2 Broadcasting Vehicle Mobility

There are many options available to a broadcaster for the type of broadcasting unit that is chosen for ENG and SNG vehicle. Some of the more important broadcasting options are shown (on similar type vehicles) in Figures 14.1, 14.2, and 14.3, and the implications of these options for portability and mobility can be clearly seen from the illustrations. The less alignment that is needed with the transmitter antenna system used to form the transmission link, the more mobile the vehicle will be.

The first vehicle shown in Figure 14.1 is a typical SNG vehicle, with a parabolic satellite dish mounted on top, which must have a clear unobstructed and sloping view to the sky. This vehicle must be configured for broadcasting at each location and the transmitter power slowly brought up to the set value (typically ~ 300W). The vehicle uses a parabolic dish with mechanical tracking, which may or may not be connected to an electronic control circuit to ensure that the dish is pointed toward the satellite of interest. Either way the tracking tends to be a slow process and is

Figure 14.1 SNG vehicle.

possible only when the shock due to motion is not severe. This means that the vehicle must be stationary and cannot move to follow events. In recent times much work has been carried out to develop flat antennas which would allow greater mobility to SNG vehicles by employing a combination of antenna beam steering techniques and mechanical tracking of the satellite using a beacon signal. These are complicated methods which can give the SNG vehicle increased signal stability while stationary or moving [1].

The second type of vehicle is indicative of an extendible mast, VHF/UHF or microwave, link contribution vehicle. This ENG vehicle is also highly portable but once at a location, it must reside there while broadcasting. The transmitter output power can be of the order of tens of watts, and the link is usually a permanent low cost asset, which is owned by the network provider. (Note that the antenna shown in Figure 14.2 is illustrative of a UHF system, whereas later in Figure 14.5 a typical microwave horn type antenna is illustrated.)

The third type of vehicle, shown in Figure 14.3, is a fixed antenna type vehicle. It, like the ENG vehicle, could use UHF/VHF or microwave frequencies to broadcast on. However it is the only class that is truly mobile as it uses a low height omnidirectional whip antenna. It can broadcast both while stationary and moving. This is the generic type of broadcasting that is receiving much attention recently due to recent developments in digital terrestrial television technology. It will be covered in more detail in this chapter, but the reader is recommended to review the chapters on MPEG-2 encoding, and COFDM modulation prior to proceeding in this chapter.

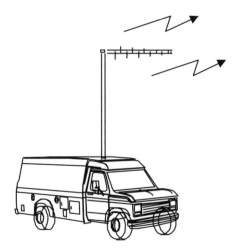

Figure 14.2   Extendible mast vehicle.

14.3   Portable and Mobile Broadcasting

There is a difference between both of these broadcasting applications. Both types are closely related, and both are used for contribution links, that is, from a particular place back to a studio before the source material is edited and processed. They are not used to broadcast material directly to the consumer however in live sporting and news events this process is nearly instantaneous and thus would appear so to the general public.

Portable broadcasting is any portable method by which a broadcast contribution can be made back to a studio. It can involve the use of satellite technology such as in SNG vehicles, or other linking equipment including telephony (for voice circuits) and also microwave circuits as used in ENG vehicles. The equipment is housed in a vehicle which can be driven to the required location, parked, and set-up. There is always a set-up time that broadcasters try to minimize and technical support is usually required. With any satellite based system there will be many locations that will not be suitable for linking purposes especially in urban areas. This is due to the requirement for a line of sight to the particular satellite being used, which can easily be obstructed by buildings and other obstacles. However in most cases a nearby alternative location can be found. Satellite transponder bandwidth costs are quite high on an annual basis, and as a result a number of SNG units can often share the same transponder bandwidth without undue resource conflict.

Figure 14.3 Fixed antenna vehicle.

Mobile broadcasting is the name given to any mobile means of broadcasting a contribution from a remote site to a studio. It could involve using a mobile phone to broadcast a voice signal on a train, or could involve broadcasting a live video signal from a moving vehicle or helicopter. Although this type of outside broadcast link (OBL) is denoted as mobile broadcasting, it also includes portable broadcasting, as portability is a necessary requirement of mobile broadcasting. Traditional mobile links have used analog microwave or UHF technology to broadcast from the mobile unit (MU) back to an elevated receive aerial, and from there into a link network and back to the studio. These analog circuits have been known to suffer from multipath ghosting from time to time as the mobile broadcast unit moved in and out of shadowed areas. However the capital and operating costs are relatively low and it has been very successful in radio traffic and news reporting.

As was outlined in the above paragraph, the capability for portable broadcasting is a necessary (but not sufficient) requirement for mobile broadcasting, and therefore any developments in mobile broadcasting techniques will improve the performance of portable units equally.

14.4 COFDM for Mobile Broadcasting

This section will focus on digital mobile broadcasting and in particular upon the implications of combining COFDM technology in conjunction with QPSK modulations and MPEG-2 encoding techniques. The combination of these innovative digital television technologies has allowed the realization of multipath resilient, power efficient, and highly compressed digital ENG links on terrestrial and microwave channels.

As was seen in earlier chapters, COFDM is a digital television (DTV) channel encoding method that has built in features to guard against long echoes due to multipath, and as such has been endorsed by ETSI and other bodies as a suitable channel encoding scheme for digital terrestrial television (DTT). However, it has also received attention as a suitable method for the transmission of contribution circuits especially for digital outside broadcasting links (OBL) [2].

In traditional live TV transmissions of sporting events such as marathons, cycling races, and also for helicopter news feeds, a moving OB unit transmits the source material using a frequency modulation (FM) method, typically in the UHF band. Due to the movement of the unit severe multipath can be noticed and the transmit aerial (if directional) must be orientated continuously to a number of receive aerials to ensure that the overall link does not fail.

Early experimental work and modelling had suggested that COFDM would be efficient in digital OBL for mobile applications [3]. However until low cost (DVB-T) COFDM modulators were developed the technique was not widely used commercially. Despite the unavailability of standardized modulators at affordable prices, results from different experiments pointed the way toward multipath free mobile broadcasting using COFDM modulation in conjunction with QPSK inner modulation [4]. Now that COFDM modems are readily available for digital terrestrial television (DTT) broadcasting, it is becoming apparent to the broadcasting community that only a very small subset of the functions of DTT modulators are being used in digital terrestrial transmissions. For instance most broadcasting organizations will only ever use 64-QAM as an inner modulation to map the MPEG-2 data signal onto the COFDM carriers. This is because the data capacity is too low with the other inner modulations (such as QPSK and 16-QAM) for economic distribution of television signals to the home. However as can be seen in Table 7.4 there are other operating modes available for selection with DVB-T, which are more rugged and would have sufficient data carrying capacity for contribution links such as mobile broadcasting. QPSK is a very rugged inner modulation that is useful in power limited transmission environments such as satellite links, and of course mobile broadcasting. QPSK is used extensively to modulate data onto single carrier systems for DSNG because of the superior carrier to noise (C/N) performance that it can achieve with low transmission powers. QPSK is also available to be used within COFDM terrestrial television modulators. As a result of recent development work by manufacturers and field trials by broadcasters, DTV modulators are now being marketed that are suited to ENG type applications.

14.4.1 Compatibility With Existing Equipment

In the future it is anticipated that more and more studio and contribution equipment will use MPEG-2 as a format for editing, processing, and contribution/distribution links. As mentioned earlier it is regarded as good practice to minimize the number of format changes between data acquisition and transmission, so it will be more attractive to use MPEG-2 encoding for the source material in the mobile unit. This will mean that the source material will also be compatible with any DVB-T compliant COFDM modulators used to broadcast back to the studio. As mentioned above terrestrial COFDM modulators are now readily available and can be easily configured for the mobile broadcasting environment.

14.4.2 Inner Modulation (QPSK)

In Chapter 7, it was shown that many different inner modulations can be chosen for mapping data onto COFDM carriers with DVB-T modems, the trade-off is between data payload and ruggedness. We have also seen that QPSK is well known for the excellent performance it has at low transmitter powers, in DSNG applications. It is possible to use therefore standard DVB-T modulators for mobile and portable broadcasts. A verification of this possibility was performed when mobile broadcasts were made using QPSK as inner modulation with a terrestrial COFDM modulator (as found in DTV COFDM transmitters) [5]. By selection of a reasonably long guard interval duration (GI = 1/8), it was possible to carry an MPEG-2 data signal at a net bit rate of 6 Mbps, without any multipath fading or ghosting effects. This was possible in urban areas despite there being no line of sight between the whip transmit aerial and the receive aerial. It is accepted that a 6 Mbps MPEG-2 payload can correspond to a contribution quality video, and associated stereo audio, signal, using modern compression techniques.

14.4.3 Spectrum Choices for Mobile Applications

Mobile broadcasting using the terrestrial COFDM signal according to the DVB-T standard requires the usage of a standard bandwidth television channel. In an era of scarce spectrum due to the introduction of more and more new analog and digital television channels, it is difficult to obtain any terrestrial frequency assignments for contribution channels. However it is possible to use practically any free spectrum in the VHF or UHF bands as well as microwave bands for mobile contribution circuits. Clearly the VHF and UHF bands are more attractive, having better propagation characteristics, and also there are usually so-called "taboo channels" available in these bands. These taboo channels are unavailable for normal broadcast usage due to the well known spectrum polluting effects of analog television, however they can be suitable for mobile digital OB links. The experiments, mentioned in the previous section, were performed with low power mobile OB links on upper adjacent channels to system PAL I, high power transmitters. The results have proven that these taboo channels work well. The rugged QPSK/COFDM signal can be received using a wide beam-width antenna, which can be cosited on the same mast as the high power transmit antenna used by the adjacent channel analog transmitter. By using these analog taboo channels, it may be possible to find suitable frequency assignments for mobile broadcasting. These channels are normally unavailable for analog broadcasting due to frequency planning criteria it is therefore a spectrum efficient means of establishing contribution circuits.

14.4.4 Possible Network Configuration

As described earlier the mobile unit could use MPEG-2 encoding in conjunction with a COFDM modulator to broadcast a robust signal from a whip type aerial to a suitable receive point. It is possible to establish a receive system at a suitable location for the contribution circuit of the mobile terrestrial broadcasting unit, using a combination of MPEG-2 network adapters and a COFDM demodulator. A network arrangement is shown in Figure 14.4, for possibly a main station transmitter site, which would be in the vicinity of the mobile unit, but remote from the required destination of the contribution circuit. The broadcast tower or mast could have a separate receive antenna system rigged to receive the desired mobile broadcast channel. After suitable channel filtering of the RF signal, which is in a digital terrestrial television format, the contribution circuit could be demodulated and the MPEG-2 TS extracted at the main-station site. This signal could then be passed to a network adapter to adapt it to a suitable telecommunications signal interface (e.g., G.703 format). After transmission to the studio within a telecommunications link network, the MPEG-2 signal could again be extracted for editing and processing. The MPEG-2 signal in this example is not encrypted and

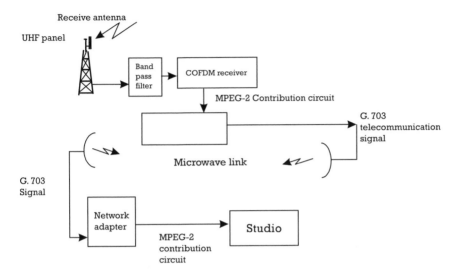

Figure 14.4 Mobile contribution circuit—possible network configuration.

therefore since an omnidirectional transmit aerial is used at the mobile unit, the signal could in principle be received by any domestic DVB-T receiver correctly tuned to the mobile broadcast channel, for the duration of the contribution.

14.5 Building Vehicles for Mobile and Portable Broadcasts

Typical broadcasting vehicles represent quite large capital expenditure for any broadcasting organization, and a vehicle building project is usually initiated to design and build these vehicles. These projects require a lot of planning, discussion, and engineering input to avoid expensive mistakes. As a result project management approaches are involved in the careful development of these vehicles. It is always recommended to spend time in the specification phase of the mobile vehicle project. It is necessary for the engineering team involved in the building of such vehicles to have a clear idea of the primary purpose of such a vehicle, whether it is intended to be used primarily for news applications, sports events, or for entertainment events. Often post production facilities may be needed in the vehicle as well as the ability to provide live contributions directly to the home studio. It is clearly important for the project team to understand the primary objectives of the vehicle, and it is also important to have an idea of the projected uses for such a capital asset, as often the needs of the organization will change and the vehicle will be required to perform more tasks than the original designer planned for. This can result in expensive additions to the vehicle, which might have been avoided with more communication during the early phases of the project.

14.5.1 Specification Phase

Early on in the building of a mobile or portable broadcasting type vehicle it is important to garner the views of experienced professionals who both may use the vehicle and/or have experience of building similar vehicles for similar applications. The designers must know the planned and potential uses of the vehicle. The type of vehicle and its size based on the planned uses can next be decided. It is important to list all the equipment that will be installed into the vehicle to get an idea of the total cost of the equipment, the physical size and weight of the equipment, and the power consumption of the equipment. This information is needed, for among other things, calculations of the expected power loading on the generator or other power supply units. With this information an initial shortlist of suitable vehicles can be made. Other things to bear in mind are the requirements for special driving qualifications for large vehicles, and technical operation capability for the operational staff within the vehicle. If a single operator is desired, then a large truck with much complexity may not be suitable.

14.5.2 Contracting Phase

After an initial specification has been agreed on it is time to begin to select a suitable contractor or contractors to build the vehicle. The purchasing organization must choose between a turnkey system where the vehicle is completely built by the contractor or it might choose an in-house project where the different parts of the project are contracted to different organizations. Either way the purchasing organization should be involved in all aspects of the contracting phase to ensure that problems are sorted as they arise. There are a number of parts to the contracting phase including:

1. Vehicle coach building—This is where the vehicle is fitted out with all of the mechanical equipment needed in a mobile broadcasting

vehicle. This will include all racks, power systems, lighting, walls, floors, seating, electric generators, extendible mast, ladders, safety interlocks, and so forth. An issue will be the weight loading on the vehicle and the contractor may be required to add extra suspension to the vehicle, and possibly add mechanical dampers under the generator and equipment racks to provide mechanical isolation for the equipment from the vehicle. A provision for climate control equipment should be made as it will be required in almost all-operating scenarios.

2. Technical equipment selection—This is where the actual broadcasting equipment is selected and purchased to meet the requirements set in the specification phase of the project. Equipment should be rack mountable and have a high degree of insensitivity to adverse climates, and mechanical shock due to bumping and movement on uneven road surfaces. The equipment may have to work at short notice or when the vehicle is in a very cold climate. If the vehicle is parked and unattended for a long period without heating the equipment will become cold and possibly damp, so it should not be overly sensitive to dampness or equally humidity.

3. Technical fitting out—This is where the equipment is actually installed into the racks and mounting bays. It also includes the wiring of signal cables, patch panels, and terminating all cables in the correct connector types, as well as labelling circuits. Engineering drawings will be needed for the contractor to follow. Vehicle commissioning will take place at this time to insure that each component of the broadcast vehicle works independently, and finally as a complete unit. Any redundancy in equipment and signal routing can be tested with equipment failure simulations. Final acceptance testing is then performed, including documentation verification.

14.5.3 Equipment Hand-Over

When the vehicle is completed it will eventually be handed over from the project staff (including the contractors and internal engineering) to the operational staff. It can be expected that training will be needed in the operation of the equipment and in the health and safety aspects of the equipment. Project staff should provide all relevant documentation to operations. It is not required for operations to have access to all of the technical manuals within the vehicle, as this would complicate matters, but operational procedures should be clearly documented and summarized fault finding techniques are recommended. Support from the project engineering staff and contractors is a requirement, especially in the early hand-over phase of the project. Modifications to the design may be required in the future and the contractor should have the interest, experience, and capability to deliver these as the need arises [6].

14.6   Mobile and Portable Microwave Link Failure

There are many reasons for the failure of any contribution link and they can happen despite the efforts of operational staff to establish secure links from the vehicle to the receive site in advance of the broadcast. Quite often a link, which was established well in advance of a live broadcast and had been working well prior to the live broadcast, can appear to fade suddenly and without reason during the sensitive live broadcast. Digital link systems unlike analog systems can fade catastrophically, because a digital system will work without degradation until the point of failure is reached. This is due to the cliff effect, described earlier (see Chapter 9). Digital encoders can be quite sensitive at times to cross-edits and cuts which are not properly synchronized. However, assuming that there is neither equipment failure nor human error, signal fades can be the most troublesome cause of failure for microwave contribution circuits.

14.6.1 Microwave Fresnel Zones

Microwave transmission is regarded as line of sight. This is because the direct path for the main signal, that is, between the transmitter antenna and the receive antenna, must have no direct obstructions otherwise the signal will be blocked. However, microwave propagation is in three dimensions and as a result there is a zone around the main signal path within which the energy of the signal is concentrated, and this zone is widest midway between the transmitter and receiver. This zone is called the Fresnel zone (or sometimes the first Fresnel zone) and generally it is better if the complete Fresnel zone has clearance from all obstructions, and not just the main path. The signals within the Fresnel zone will add in the receiver without signal degradation irrespective of signal path. The envelope containing these paths is called the Fresnel zone, and is defined as the envelope containing all paths with a length equal to the direct distance between the transmitter and receiver plus 1/2 the wavelength at the transmitter frequency. For normal operation designers recommend using a clearance of 0.6 the Fresnel zone. This is because the received signal energy drops off significantly above and below the first Fresnel zone. It is accepted that 0.6 Fresnel zone clearance permits normal atmospheric variations to occur without destructive losses [7].

14.6.2 Building Reflections
Buildings and other large obstructions can cause reflections of microwave signals. Paths looking down city streets are regarded as the most severe for reflections due to the flat building surfaces. With analog microwave links these reflections could cause total signal destruction if the receive signals were out of phase. However with some modern digital transmission schemes, the reflected signals will add constructively and augment the receive signal. However, it is good practice to avoid these paths where possible.

14.6.3 Refraction of Microwave Signals
The refraction of electromagnetic energy, such as light and RF signals as they pass through the boundaries of other transmission media, is a well-known phenomenon. Broadcasters have observed microwave refraction, that is, the bending of microwave signals as they pass through the air for many years, often to their dismay! The temperature of air is known to change from layer to layer within the atmosphere, and as a result the density of the air also changes, and this causes a change in the refractive index of this medium. The refractive index of air is inversely proportional to its density, so as air density decreases the refractive index will increase. This unwanted refraction of microwave signals in the atmosphere is similar in principle to the refractive nature of light energy within media such as glass, which in this case is exploited as a beneficial property and is used as the basis of fiber-optic transmission.

14.6.4 Thermoclines and Inversion Layers
Within the atmosphere different layers of air which are at different temperatures can be created quickly. These layers are naturally occurring and can be created at different times of the day. They can cause different propagation properties for RF and microwave signals due to the different refractive indices of the layers. One of the problems for broadcasters is that these layers can be produced or changed within minutes. The most important atmospheric conditions for microwave propagation include the

formation of thermoclines and inversion layers. Thermoclines are narrow boundaries between two air layers that are at different temperatures. Inversion layers are thermoclines that are formed due to the cooling of air layers at different rates. These layers of warm and cold air can cause a microwave signal incident on the layer boundaries to be refracted toward the ground or away from the receive antenna, and either effect can cause a signal to fade significantly or even result in complete signal loss. These can affect MMDS systems as well as microwave ENG systems.

14.6.4.1 Typical Thermoclines
The most commonly reported cause of thermoclines is the proximity to cities and large bodies of water [8]. In winter cities tend to be warmer than the surrounding suburbs and countryside. This will cause a layer of warm air to shroud the city center. Large bodies of water such as the sea and large lakes heat up and cool at a slower rate than the earth. This causes the air on water and on land to be at different temperatures, with different refractive indices. Every night at sunset the temperature of the land drops and if the climatic conditions are right inversion layers can be formed close to the ground. All of these conditions can cause the signal level to fade by as much as 20 dB at a receiver within a short period of time.

14.6.5 Some Preventative Measures
There are a number of preventative measures that ENG vehicle operators can employ to reduce the risk of signal loss due to inversion layer formation. These are based upon a knowledge of local weather patterns and terrain topography, and how they can effect ENG transmission using microwave signals. These rules of thumb are loosely based upon operators' experiences and are intended to act as a guide only.

The easiest way to prevent fades due to localized climatic changes such as the inversion layers just described is to have some degree of path diversity. Another vehicle or another transmission method may provide the required back up. However this is the most costly preventive measure. It may be easier to have a number of possible receive sites that could be selected by either the studio or vehicle to broadcast to/from.

Other measures include locating the microwave transmitter and receiver within the same altitude band if possible. This will reduce the number of potential thermal layers that could exist between the transmitter and receiver, as shown in Figure 14.5 below. In this example three thermal layers are shown, each with a different temperature, where T1

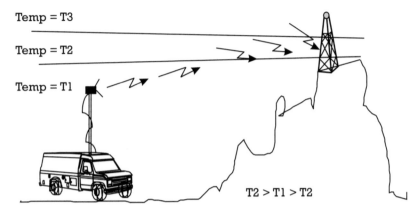

Figure 14.5   Thermal layers and microwave signal refraction.

corresponds to the ground air temperature, T2 is the inversion layer, and T3 is the typical air temperature at the receive location. Here the signal is refracted at the boundaries of the inversion layer.

Another way to increase the reliability of the path is to minimize the path length as much as possible. This will not always be practicable, however by minimizing the distance from transmitter to receiver it is possible to reduce the risk of signal loss due to fades.

Finally the risk of fade can be avoided if the transmitter and receiver are placed on the same side of the actual thermal layer. This is because the signal propagation is only affected when passing through different thermal layers. The signal will not be refracted if it does not pass through a layer boundary. Movement of the vehicle away from a large heat source or heat sink can have a marked improvement on signal propagation. For instance, a large body of water can act as source of thermal layers and these layers can adversely affect broadcasts at certain times of the day. By simply moving a little further away from the source of the layers it may be possible to reduce the risk of signal fades significantly.

References

[1]   Mitsumoto, H., et al, "A Mobile Satellite News Gathering System Using a Flat Antenna," *IEEE Trans. On Broadcasting*, Vol. 42, No. 3, September, 1996, pp. 272–277.

[2]   Matthews, N., "The ENG/OB Link," *World Broadcast News*, November, 1998, pp. 60–62.

[3]   Moriyama, S., and K. Tsuchida, "A Study on Field Pickup Unit Using OFDM Modulation Scheme," *ITE Tech. Rep*, Vol. 19, No. 38, 1995, pp. 7–12.

[4]   Moriyama, S., et al, "Digital Outside Broadcasting Link Using OFDM Modulation Scheme," *IEEE Trans. On Broadcasting*, Vol. 42, No. 3, September 1996, pp. 266–271.

[5]   O'Leary, S., D. Priestly, and N. McSparron, "Mobile Broadcasting of DVB-T Signals," *IEEE Trans. On Broadcasting*, Vol. 44, No. 3, September, 1998, pp. 346–352.

[6]   Bennett, B., "Going Mobile," *Broadcast Engineering*, January, 1999, pp. 90–95.

[7]   Tesh, K., "Terrain Factors Part 2, Further Consideration of Propagation Factors," *Wireless Broadcasting Magazine*, November, 1994, pp. 38–40.

[8]   Schaut, G., "Understanding ENG Link Failures," *Broadcast Engineering*, January 1999, pp. 96–99.

Glossary

AAC	Advanced Audio Coding standard
ACI	Adjacent channel interference
ASI	asynchronous serial interface
AES	Audio Engineering Society
A/D	analog to digital converter
AM	amplitude modulation
AM-VSB	amplitude modulated vestigial sideband
ANSI	American National Standards Institute
ATSC	Advanced Television System Committee
application	This is interpretable code which is stored within the set top box memory, either in ROM or FLASH, or received into the set top box to allow the user to interact with services.
API	application programming interface
application streaming	An application based upon a stream of data that is continuously broadcast, such as weather information, or stock market information.
Aural carrier	An analog audio signal at a broadcast frequency.
BBC	British Broadcasting Corporation

BER	bit error rate
BJT	bipolar junction transistors
BW	bandwidth
bouquet	A collection of services marketed as a single entity.
CATV	cable antenna television
CCI	co-channel interference
CCITT	Comite Consultatif Internationale Telegraphe et Telephone (an ITU organization)
CENELEC	Comite Europeen de Normalisation Electrotechnique
CI	common interface
conditional access	Any method which limits access to a service or application to a group of authorized users only.
COFDM	coded orthogonal frequency division multiplexing
CIF	common interleaved frames
CBR	constant bit rate
C/N	Carrier-to-noise ratio
CPE	common phase error
CU	capacity units
DAB	digital audio broadcasting
DAVIC	Digital Audio-Visual Council
dB	decibels
dBi	decibels relative to an isotropic radiator
DBS	direct broadcast satellite
DCT	discrete cosine transform
DSP	digital signal processing
DTS	decoding time stamp
Digitag	Digital Terrestrial Television Action Group

diplexer	A device that combines two signals
DTG	U.K. Digital TV group
DTH	direct to home
DiBEG	Digital Broadcasting Experts Group
DIAMOND	digital interactive multimedia on demand
D/A	digital to analog converter
DAB	digital audio broadcasting
DEMUX	demultiplexer
DPCM	differential pulse code modulation
downloading	This is a process whereby an application, service, or movie is downloaded to the set top box over the network for consumption by the user.
DQPSK	differential-QPSK
DRM	Digital Radio Mondiale
DSMCC	digital storage media control and command
DSNG	digital satellite news gathering
DSTB	digital set top box
DTV	digital television
DTT	digital terrestrial television
DVB	The Digital Video Broadcasting project
DVB-C	DVB cable specification
DVB-CI	DVB common interface specification
DVB-MHP	DVB multimedia home platform
DVB-S	DVB satellite specification
DVB-SI	DVB service information
DVB-T	DVB terrestrial specification
DVB-TRC	DVB terrestrial return channel specification

EBU	European Broadcasting Union
e-commerce	electronic commerce
e-mail	electronic mail
EDTV	extended definition television
EIT	event information table
EN	European norm
EPG	electronic program guide
ERP	emitted radiated power
ES	elementary streams
ETI	ensemble transport interface
ETR	European technical requirements
ETS	European Technical Standard
ETSI	European Telecommunications Standards Institute
EU	European Union
encryption	A coding system used to prevent unauthorized users from decoding a service and gaining access to an application.
ELG	European Launching Group
ENG	electronic news gathering
electronic program guide	This is an application to allow the user to navigate through the various services on offer. It also provides information to the user about programs and services in an electronic format, similar to a television guide. It can be overlaid on program images.
FEC	forward error correction
FET	field effect transistor
FFT	Fast Fourier Transformation
FCC	Federal Communication Commission

FDM   Frequency Division Multiplexing

FM   frequency modulated

field   The intensity of an electromagnetic wave, measured in
strength   volts per meter.

firmware   The software layer within the set top box used to drive
                 devices such as smart card readers and return channel
                 modems.

FSL   free space loss (path loss)

Fresnel zone A circular band of energy surrounding a propagated sig-
                 nal defined by the difference in path length between
                 boundaries of the circles.

GA   Grand Alliance

GB   guard band

GOP   group of pictures

guard   A period of time when all delayed signal components are
interval   constructively added in a receiver.

graphical   A visual representation of an electronic application,
user interface   which the user interacts with to engage an application.
(GUI)

graceful   A gradual reduction in field strength and signal quality.
degradation

GPS   global positioning system

GSM   Group Speciale Mobile

HDTV   high definition television

HP   high priority

HTML   Hypertext mark-up language is a computer mark-up lan-
                 guage used to create hypertext and hypermedia docu-
                 ments on the World Wide Web. It can incorporate text,
                 graphics, sound video, and hyperlinks.

IBOC   in band on channel

ICI   intercarrier interference

IDTV	interactive digital television
IFFT	Inverse Fast Fourier Transform
IF	intermediate frequency (typically 35 MHz, or 70 MHz)
IOT	inductive output tube
ITU	International Telecommunication Union
IEC	International Electrotechnical Commission
ISO	International Standards Organization
ISDB-T	integrated services digital broadcasting—terrestrial
IP	Internet protocol
IRD	integrated receiver decoder
Java	An object-orientated programming language that is suited to programming in networked environments. It was developed by Sun Microsystems and is used in DTV API programs.
JTC1	Joint Technical Committee 1
LVDS	low voltage differential signal
LP	low priority
LW	long wave
MER	modulation error rate
MTBF	mean time between failure
MW	medium wave
middleware	This is software contained within the digital set top box between the applications and the firmware.
MIP	megaframe initialization packet
MAC	media access control
MAC	multiplexed analog component
Metadata	data describing other data
MHP	multimedia home platform

MHEG	Multimedia and Hypermedia Expert Group
MFN	multifrequency network
MPEG	Moving Pictures Expert Group
MP @ ML	main profile at main level
MOT	multimedia object transfer standard
MSC	main service channel
MUX	multiplexer
Multiplex	An MPEG-2 transport stream, a combination of digital services to be broadcast on a single broadcast channel.
MUSICAM	masking pattern adapted universal sub-band integrated coding and multiplexing
MMDS	microwave multipoint distribution system
MVDS	microwave video distribution system
NA	network adapted
NI	network independent
NIT	network information table
NICAM	near instantaneous companding audio multiplex
NTSC	National Television Systems Committee
OB	outside broadcast
OBU	outside broadcast unit
OFDM	orthogonal frequency division multiplexing
OBO	output power backoff
PAL	phase alternate line
PAT	program association table
PES	packetized MPEG-2 elementary streams
PC	personal computer
PC Card	See PCMCIA

PCMCIA   Acronym for Personal Computer Memory Card Interna-
                    tional Association. It is a standard interface used with
                    insertable cards pushed into a computer or set top box
                    (also known as PC card).

pixels   picture elements

protection   The minimum value of the wanted to unwanted signal
ratio   ratio.

PCM   pulse code modulation

PCR   program clock reference

PDH   plesiochronous digital hierarchy

PPS   pulse per second

PRBS   pseudo random binary sequence

prETS   pre-European Technical Standard

PS   program stream

PSI   program Specific Information

PSK   phase shift keying

QAM   quadrature amplitude modulation

QEF   quasi-error free

QPSK   quadrature phase shift keying

QoS   Quality of Service

RF   radio frequency

RS   Reed Solomon

RLC   run length coding

RCPC   rate compatible punctured convolutional code

SDH   synchronous digital hierarchy

SDTV   standard definition television

SECAM   systeme electronique couleur avec memoire

SI   service information

SIIP	service information insertion points
SFN	single frequency network
SPECTRE	Experimental European Research Program
SW	short wave
SMATV	satellite master antenna television
smart card	This is a credit card sized plastic card that contains an electronic processor chip. It is inserted into an STB to obtain access rights to services or applications. The information contained on the chip can be updated remotely via the network delivery medium.
SNR	signal-to-noise ratio
STB	set top box
STU	set top unit
stream	A flow of data. An MPEG-2 transport stream is a flow of data that contains possibly video images together with audio signals, and data.
SPI	synchronous parallel interface
TDT	time and date table
TOT	time offset table
TIE	transmitter input equipment
TPS	transmission parameter signalling
TS	transport stream
TV	television
UHF	ultra high frequency

virtual machine	This is a software system which converts programs into simple binary instructions, similar to machine code. These instructions are in a universal format that is not specific to a particular processor. All the normal activities of a real microprocessor can be performed in a safe and virtual environment using this system.
VBI	vertical blanking interval
VSB	vestigial sideband modulation
VHF	very high frequency
VLC	variable length coding
VLSI	very large scale integration
VBI	vertical blanking interval
UTC	universal coordinated time
visual carrier	A transmitter output signal at the channel frequency, and normally containing the analog visual and color carriers.

About the Author

Seamus O'Leary holds a degree in electronic engineering (B.E.) from University College Dublin, and a certified diploma in accountancy and finance (C.Dip.A.F.) from the Chartered Association of Certified Accountants (ACCA). He is a chartered engineer (C.Eng.) and works as a project manager in the network business unit capital projects of RTE, the Irish national broadcaster. Before joining RTE, he was technical manager of North West Labs Ltd., a telecommunications consultancy company. He has worked on various projects associated with analog and digital television and radio for the past six years, and before that worked on high-speed optical networks and equipment. He has written many papers on digital transmission, and he has won both the Mullins silver medal and the Smith testimonial award from the Institution of Engineers of Ireland for his work in this field. He is a member of IEEE.

Index